CW01500334

METEOROLO...
BY

Aristotle

Translated by Theodorus Gaza

Meteorology

Book I

Part 1

WE have already discussed the first causes of nature, and all natural motion, also the stars ordered in the motion of the heavens, and the physical element-enumerating and specifying them and showing how they change into one another-and becoming and perishing in general. There remains for consideration a part of this inquiry which all our predecessors called meteorology. It is concerned with events that are natural, though their order is less perfect than that of the first of the elements of bodies. They take place in the region nearest to the motion of the stars. Such are the Milky Way, and comets, and the movements of meteors. It studies also all the affections we may call common to air and water, and the kinds and parts of the earth and the affections of its parts. These throw light on the causes of winds and earthquakes and all the consequences the motions of these kinds and parts involve. Of these things some puzzle us, while others admit of explanation in some degree. Further, the inquiry is concerned with the falling of thunderbolts and with whirlwinds and fire-winds, and further, the recurrent affections produced in these same bodies by concretion. When the inquiry into these matters is concluded let us consider what account we can give, in accordance with the method we have followed, of animals and plants, both generally and in detail. When that has been done we may say that the whole of our original undertaking will have been carried out.

After this introduction let us begin by discussing our immediate subject.

Part 2

We have already laid down that there is one physical element which makes up the system of the bodies that move in a circle, and besides this four bodies owing their existence to the four principles, the motion of these latter bodies being of two kinds: either from the centre or to the centre. These four bodies are fire, air, water, earth. Fire occupies the highest place among them all, earth the lowest, and two elements correspond to these in their relation to one another, air being nearest to fire, water to earth. The whole world surrounding the earth, then, the affections of which are our subject, is made up of these bodies. This world necessarily has a certain continuity with the upper motions: consequently all its power and order is derived from them. (For the originating principle of all motion is the first cause. Besides, that element is eternal and its motion has no limit in space, but is always complete; whereas all these other bodies have separate regions which limit one another.) So we must treat fire and earth and the elements like them as the material causes of the events in this world (meaning by material what is subject and is affected), but must assign causality in the sense of the originating principle of motion to the influence of the eternally moving bodies.

Part 3

Let us first recall our original principles and the distinctions already drawn and then explain the 'Milky Way' and comets and the other phenomena akin to these.

Fire, air, water, earth, we assert, originate from one another, and each of them exists potentially in each, as all things do that can be resolved into a common and ultimate substrate.

The first difficulty is raised by what is called the air. What are we to take its nature to be in the world surrounding the earth? And what is its position relatively to the other physical elements. (For there is no question as to the relation of the bulk of the earth to the size of the bodies which exist around it, since astronomical demonstrations have by this time proved to us that it is actually far smaller than some individual stars. As for the water, it is not observed to exist collectively and separately, nor can it do so apart from that volume of it which has its seat about the earth: the sea, that is, and rivers, which we can see, and any subterranean water that may be hidden from our observation.) The question is really about that which lies between the earth and the nearest stars. Are we to consider it to be one kind of body or more than one? And if more than one, how many are there and what are the bounds of their regions?

We have already described and characterized the first element, and explained that the whole world of the upper motions is full of that body.

This is an opinion we are not alone in holding: it appears to be an old assumption and one which men have held in the past, for the word ether has long been used to denote that element. Anaxagoras, it is true, seems to me to think that the word means the same as fire. For he thought that the upper regions were full of fire, and that men referred to those regions when they spoke of ether. In the latter point he was right, for men seem to have assumed that a body that was eternally in motion was also divine in nature; and, as such a body was different from any of the terrestrial elements, they determined to call it 'ether'.

For the um opinions appear in cycles among men not once nor twice, but infinitely often.

Now there are some who maintain that not only the bodies in motion but that which contains them is pure fire, and the interval between the earth and the stars air: but if they had considered what is now satisfactorily established by mathematics, they might have given up this puerile opinion. For it is altogether childish to suppose that the moving bodies are all of them of a small size, because they so to us, looking at them from the earth.

This is a matter which we have already discussed in our treatment of the upper region, but we may return to the point now.

If the intervals were full of fire and the bodies consisted of fire every one of the other elements would long ago have vanished.

However, they cannot simply be said to be full of air either; for even if there were two elements to fill the space between the earth and the heavens, the air would far exceed the quantity required to maintain its proper proportion to the other elements. For the bulk of the earth (which includes the whole volume of water) is infinitesimal in comparison with the whole world that surrounds it. Now we find that the excess in volume is not proportionately great where water dissolves into air or air into fire. Whereas the proportion between any given small quantity of water and the air that is generated from it ought to hold good between the total amount of air and the total amount of water. Nor does it make any difference if any one denies that the elements originate from one another, but asserts that they are equal in power. For on this view it is certain amounts of each that are equal in power, just as would be the case if they actually originated from one another.

So it is clear that neither air nor fire alone fills the intermediate space.

It remains to explain, after a preliminary discussion of difficulties, the relation of the two elements air and fire to the position of the first element, and the reason why the stars in the upper region impart heat to the earth and its neighbourhood. Let us first treat of the air, as we proposed, and then go on to these questions.

Since water is generated from air, and air from water, why are clouds not formed in the upper air? They ought to form there the more, the further from the earth and the colder that region is. For it is neither appreciably near to the heat of the stars, nor to the rays reflected from the earth. It is these that dissolve any formation by their heat and so prevent clouds from forming near the earth. For clouds gather at the point where the reflected rays disperse in the infinity of space and are lost. To explain this we must suppose either that it is not all air which water is generated, or, if it is produced from all air alike, that what immediately surrounds the earth is not mere air, but a sort of vapour, and that its vaporous nature is the reason why it condenses back to water again. But if the whole of that vast region is vapour, the amount of air and of water will be disproportionately great. For the spaces left by the heavenly bodies must be filled by some element. This cannot be fire, for then all the rest would have been dried up. Consequently, what fills it must be air and the water that surrounds the whole earth-vapour being water dissolved.

After this exposition of the difficulties involved, let us go on to lay down the truth, with a view at once to what follows and to what has already been said. The upper region as far as the moon we affirm to consist of a body distinct both from fire and from air, but varying degree of purity and in kind, especially towards its limit on the side of the air, and of the world surrounding the earth. Now the circular motion of the first element and of the bodies it contains dissolves, and inflames by its motion, whatever part of the lower world is nearest to it, and so generates heat. From another point of view we may look at the motion as follows. The body that lies below the circular motion of the heavens is, in a sort, matter, and is potentially hot, cold, dry, moist, and possessed of whatever other qualities are derived from these. But it actually acquires or retains one of these in virtue of motion or rest, the cause and principle of which has already been explained. So at the centre and round it we get earth and water, the heaviest and coldest elements, by themselves; round them and contiguous with them, air and what we commonly call fire. It is not really fire, for fire is an excess of heat and a sort of ebullition; but in reality, of what we call air, the part surrounding the earth is moist and warm, because it contains both vapour and a dry exhalation from the earth. But the next part, above that, is warm and dry. For vapour is naturally moist and cold, but the exhalation warm and dry; and vapour is potentially like water, the exhalation potentially like fire. So we must take the reason why clouds are not formed in the upper region to be this: that it is filled not with mere air but rather with a sort of fire.

However, it may well be that the formation of clouds in that upper region is also prevented by the circular motion. For the air round the earth is necessarily all of it in motion, except that which is cut off inside the circumference which makes the earth a complete sphere. In the case of winds it is actually observable that they originate in marshy districts of the earth; and they do not seem to blow above the level of the highest mountains. It is the revolution of the heaven which carries the air with it and causes its circular motion, fire being continuous with the upper element and air with fire. Thus its motion is a second reason why that air is not condensed into water.

But whenever a particle of air grows heavy, the warmth in it is squeezed out into the upper region and it sinks, and other particles in turn are carried up together with the fiery exhalation. Thus the one region is always full of air and the other of fire, and each of them is perpetually in a state of change.

So much to explain why clouds are not formed and why the air is not condensed into water, and what account must be given of the space between the stars and the earth, and what is the body that fills it.

As for the heat derived from the sun, the right place for a special and scientific account of it is in the treatise about sense, since heat is an affection of sense, but we may now explain how it can be produced by the heavenly bodies which are not themselves hot.

We see that motion is able to dissolve and inflame the air; indeed, moving bodies are often actually found to melt. Now the sun's motion alone is sufficient to account for the origin of terrestrial warmth and heat. For a motion that is to have this effect must be rapid and near, and that of the stars is rapid but distant, while that of the moon is near but slow, whereas the sun's motion combines both conditions in a sufficient degree. That most heat should be generated where the sun is present is easy to understand if we consider the analogy of terrestrial phenomena, for here, too, it is the air that is nearest to a thing in rapid motion which is heated most. This is just what we should expect, as it is the nearest air that is most dissolved by the motion of a solid body.

This then is one reason why heat reaches our world. Another is that the fire surrounding the air is often scattered by the motion of the heavens and driven downwards in spite of itself.

Shooting-stars further suffix to prove that the celestial sphere is not hot or fiery: for they do not occur in that upper region but below: yet the more and the faster a thing moves, the more apt it is to take fire. Besides, the sun, which most of all the stars is considered to be hot, is really white and not fiery in colour.

Part 4

Having determined these principles let us explain the cause of the appearance in the sky of burning flames and of shooting-stars, and of 'torches', and 'goats', as some people call them. All these phenomena are one and the same thing, and are due to the same cause, the difference between them being one of degree.

The explanation of these and many other phenomena is this. When the sun warms the earth the evaporation which takes place is necessarily of two kinds, not of one only as some think. One kind is rather of the nature of vapour, the other of the nature of a windy exhalation. That which rises from the moisture contained in the earth and on its surface is vapour, while that rising from the earth itself, which is dry, is like smoke. Of these the windy exhalation, being warm, rises above the moister vapour, which is heavy and sinks below the other. Hence the world surrounding the earth is ordered as follows. First below the circular motion comes the warm and dry element, which we call fire, for there is no word fully adequate to every state of the humid evaporation: but we must use this terminology since this element is the most inflammable of all bodies. Below this comes air. We must think of what we just called fire as being spread round the terrestrial sphere on the outside like a kind of fuel, so that a little motion often makes it burst into flame just as smoke does: for flame is the ebullition of a dry exhalation. So whenever the circular motion stirs this stuff up in any way, it catches fire at the point at which it is most inflammable. The result differs according to the disposition and quantity of the combustible material. If this is broad and long, we often see a flame burning as in a field of stubble: if it burns lengthwise only, we see what are called 'torches' and 'goats' and shooting-stars. Now when the inflammable material is longer than it is broad sometimes it seems to throw off sparks as it burns. (This happens because matter catches fire at the sides in small portions but continuously with the main body.) Then it is called a 'goat'. When this does not happen it is a 'torch'. But if the whole length of the exhalation is scattered in small parts and in many directions and in breadth and depth alike, we get what are called shooting-stars.

The cause of these shooting-stars is sometimes the motion which ignites the exhalation. At other times the air is condensed by cold and squeezes out and ejects the hot element; making their motion look more like that of a thing thrown than like a running fire. For the question might be raised whether the 'shooting' of a 'star' is the same thing as when you put an exhalation below a lamp and it lights the lower lamp from the flame above. For here too the flame passes wonderfully quickly and looks like a thing thrown, and not as if one thing after another caught fire. Or is a 'star' when it 'shoots' a single body that is thrown? Apparently both cases occur: sometimes it is like the flame from the lamp and sometimes bodies are projected by being squeezed out (like fruit stones from one's fingers) and so are seen to fall into the sea and on the dry land, both by night and by day when the sky is clear. They are thrown downwards because the condensation which propels them inclines downwards. Thunderbolts fall downwards for the same reason: their origin is never combustion but ejection under pressure, since naturally all heat tends upwards.

When the phenomenon is formed in the upper region it is due to the combustion of the exhalation. When it takes place at a lower level it is due to the ejection of the exhalation by the condensing and cooling of the moister evaporation: for this latter as it condenses and inclines downward contracts, and thrusts out the hot element and causes it to be thrown downwards. The motion is upwards or downwards or sideways according to the way in which the evaporation lies, and its disposition in respect of breadth and depth. In most cases the direction is sideways because two motions are involved, a compulsory motion downwards and a natural motion upwards, and under these circumstances an object always moves obliquely. Hence the motion of 'shooting-stars' is generally oblique.

So the material cause of all these phenomena is the exhalation, the efficient cause sometimes the upper motion, sometimes the contraction and condensation of the air. Further, all these things happen below the moon. This is shown by their apparent speed, which is equal to that of things thrown by us; for it is because they are close to us, that these latter seem far to exceed in speed the stars, the sun, and the moon.

Part 5

Sometimes on a fine night we see a variety of appearances that form in the sky: 'chasms' for instance and 'trenches' and blood-red colours. These, too, have the same cause. For we have seen that the upper air condenses into an inflammable condition and that the combustion sometimes takes on the appearance of a burning flame, sometimes that of moving torches and stars. So it is not surprising that this same air when condensing should assume a variety of colours. For a weak light shining through a dense air, and the air when it acts as a mirror, will cause all kinds of colours to appear, but especially crimson and purple. For these colours generally appear when fire-colour and white are combined by superposition. Thus on a hot day, or through a smoky, medium, the stars when they rise and set look crimson. The light will also create colours by reflection when the mirror is such as to reflect colour only and not shape.

These appearances do not persist long, because the condensation of the air is transient.

'Chasms' get their appearance of depth from light breaking out of a dark blue or black mass of air. When the process of condensation goes further in such a case we often find 'torches' ejected. When the 'chasm' contracts it presents the appearance of a 'trench'.

In general, white in contrast with black creates a variety of colours; like flame, for instance, through a medium of smoke. But by day the sun obscures them, and, with the exception of crimson, the colours are not seen at night because they are dark.

These then must be taken to be the causes of 'shooting-stars' and the phenomena of combustion and also of the other transient appearances of this kind.

Part 6

Let us go on to explain the nature of comets and the 'Milky Way', after a preliminary discussion of the views of others.

Anaxagoras and Democritus declare that comets are a conjunction of the planets approaching one another and so appearing to touch one another.

Some of the Italians called Pythagoreans say that the comet is one of the planets, but that it appears at great intervals of time and only rises a little above the horizon. This is the case with Mercury too; because it only rises a little above the horizon it often fails to be seen and consequently appears at great intervals of time.

A view like theirs was also expressed by Hippocrates of Chios and his pupil Aeschylus. Only they say that the tail does not belong to the comet itself, but is occasionally assumed by it on its course in certain situations, when our sight is reflected to the sun from the moisture attracted by the comet. It appears at greater intervals than the other stars because it is slowest to get clear of the sun and has been left behind by the sun to the extent of the whole of its circle before it reappears at the same point. It gets clear of the sun both towards the north and towards the south. In the space between the tropics it does not draw water to itself because that region is dried up by the sun on its course. When it moves towards the south it has no lack of the necessary moisture, but because the segment of its circle which is above the horizon is small, and that below it many times as large, it is impossible for the sun to be reflected to our sight, either when it approaches the southern tropic, or at the summer solstice. Hence in these regions it does not develop a tail at all. But when it is visible in the north it assumes a tail because the arc above the horizon is large and that below it small. For under these circumstances there is nothing to prevent our vision from being reflected to the sun.

These views involve impossibilities, some of which are common to all of them, while others are peculiar to some only.

This is the case, first, with those who say that the comet is one of the planets. For all the planets appear in the circle of the zodiac, whereas many comets have been seen outside that circle. Again more comets than one have often appeared simultaneously. Besides, if their tail is due to reflection, as Aeschylus and Hippocrates say, this planet ought sometimes to be visible without a tail since, as they it does not possess a tail in every place in which it appears. But, as a matter of fact, no planet has been observed besides the five. And all of them are often visible above the horizon together at the same time. Further, comets are often found to appear, as well when all the planets are visible as when some are not, but are obscured by the neighbourhood of the sun. Moreover the statement that a comet only appears in the north, with the sun at the summer solstice, is not true either. The great comet which appeared at the time of the earthquake in Achaea and the tidal wave rose due west; and many have been known to appear in the south. Again in the archonship of Euclees, son of Molon, at Athens there appeared a comet in the north in the month Gamelion, the sun being about the winter solstice. Yet they themselves admit that reflection over so great a space is an impossibility.

An objection that tells equally against those who hold this theory and those who say that comets are a coalescence of the planets is, first, the fact that some of the fixed stars too get a tail. For this we must not only accept the authority of the Egyptians who assert it, but we have ourselves observed the fact. For a star in the thigh of the Dog had a tail, though a faint one. If you fixed your sight on it its light was dim, but if you just glanced at it, it appeared brighter. Besides, all the comets that have been seen in our day have vanished without setting, gradually fading away above the horizon; and they have not left behind them either one or more stars. For instance the great comet we mentioned before appeared to the west in winter in frosty weather when the sky was clear, in the archonship of Asteius. On the first day it set before the sun and was then not seen. On the next day it was seen, being ever so little behind the sun and immediately setting. But its light extended over a third part of the sky like a leap, so that people called it a 'path'. This comet receded as far as Orion's belt and there dissolved. Democritus however, insists upon the truth of his view and affirms that certain stars have been seen when comets dissolve. But on his theory this ought not to occur occasionally but always. Besides, the Egyptians affirm that conjunctions of the planets with one another, and with the fixed stars, take place, and we have ourselves observed Jupiter coinciding with one of the stars in the Twins and hiding it, and yet no comet was formed. Further, we can also give a rational proof of our point. It is true that some stars seem to be bigger than others, yet each one by itself looks indivisible. Consequently, just as, if they really had been indivisible, their conjunction could not have created any greater magnitude, so now that they are not in fact indivisible but look as if they were, their conjunction will not make them look any bigger.

Enough has been said, without further argument, to show that the causes brought forward to explain comets are false.

Part 7

We consider a satisfactory explanation of phenomena inaccessible to observation to have been given when our account of them is free from impossibilities. The observations before us suggest the following account of the phenomena we are now considering. We know that the dry and warm exhalation is the outermost part of the terrestrial world which falls below the circular motion. It, and a great part of the air that is continuous with it below, is carried round the earth by the motion of the circular revolution. In the course of this motion it often ignites wherever it may happen to be of the right consistency, and this we maintain to be the cause of the 'shooting' of scattered 'stars'. We may say, then, that a comet is formed when the upper motion introduces into a gathering of this kind a fiery principle not of such excessive strength as to burn up much of the material quickly, nor so weak as soon to be extinguished, but stronger and capable of burning up much material, and when exhalation of the right consistency rises from below and meets it. The kind of comet varies according to the shape which the exhalation happens to take. If it is diffused equally on every side the star is said to be fringed, if it stretches out in one direction it is called bearded. We have seen that when a fiery principle of this kind moves we seem to have a shooting-star: similarly when it stands still we seem to have a star standing still. We may compare these phenomena to a heap or mass of chaff into which a torch is thrust, or a spark thrown. That is what a shooting-star is like. The fuel is so inflammable that the fire runs through it quickly in a line. Now if this fire were to persist instead of running through the fuel and perishing away, its course through the fuel would stop at the point where the latter was densest, and then the whole might begin to move. Such is a comet-like a shooting-star that contains its beginning and end in itself.

When the matter begins to gather in the lower region independently the comet appears by itself. But when the exhalation is constituted by one of the fixed stars or the planets, owing to their motion, one of them becomes a comet. The fringe is not close to the stars themselves. Just as haloes appear to follow the sun and the moon as they move, and encircle them, when the air is dense enough for them to form along under the sun's course, so too the fringe. It stands in the relation of a halo to the stars, except that the colour of the halo is due to reflection, whereas in the case of comets the colour is something that appears actually on them.

Now when this matter gathers in relation to a star the comet necessarily appears to follow the same course as the star. But when the comet is formed independently it falls behind the motion of the universe, like the rest of the terrestrial world. It is this fact, that a comet often forms independently, indeed oftener than round one of the regular stars, that makes it impossible to maintain that a comet is a sort of reflection, not indeed, as Hippocrates and his school say, to the sun, but to the very star it is alleged to accompany-in fact, a kind of halo in the pure fuel of fire.

As for the halo we shall explain its cause later.

The fact that comets when frequent foreshadow wind and drought must be taken as an indication of their fiery constitution. For their origin is plainly due to the plentiful supply of that secretion. Hence the air is necessarily drier and the moist evaporation is so dissolved and dissipated by the quantity of the hot exhalation as not readily to condense into water.-But this phenomenon too shall be explained more clearly later when the time comes to speak of the winds.-So when there are many comets and they are dense, it is as we say, and the years are clearly dry and windy. When they are fewer and fainter this effect does not appear in the same degree, though as a rule the is found to be excessive either in duration or strength. For instance when the stone at Aegospotami fell out of the air-it had been carried up by a wind and fell down in the daytime-then too a comet happened to have appeared in the west. And at the time of the great comet the winter was dry and north winds prevailed, and the wave was due to an opposition of winds. For in the gulf a north wind blew and outside it a violent south wind. Again in the archonship of Nicomachus a comet appeared for a few days about the equinoctial circle (this one had not risen in the west), and simultaneously with it there happened the storm at Corinth.

That there are few comets and that they appear rarely and outside the tropic circles more than within them is due to the motion of the sun and the stars. For this motion does not only cause the hot principle to be secreted but also dissolves it when it is gathering. But the chief reason is that most of this stuff collects in the region of the Milky Way.

Part 8

Let us now explain the origin, cause, and nature of the Milky Way. And here too let us begin by discussing the statements of others on the subject.

(1) Of the so-called Pythagoreans some say that this is the path of one of the stars that fell from heaven at the time of Phaethon's downfall. Others say that the sun used once to move in this circle and that this region was scorched or met with some other affection of this kind, because of the sun and its motion.

But it is absurd not to see that if this were the reason the circle of the Zodiac ought to be affected in the same way, and indeed more so than that of the Milky Way, since not the sun only but all the planets move in it. We can see the whole of this circle (half of it being visible at any time of the night), but it shows no signs of any such affection except where a part of it touches the circle of the Milky Way.

(2) Anaxagoras, Democritus, and their schools say that the Milky Way is the light of certain stars. For, they say, when the sun passes below the earth some of the stars are hidden from it. Now the light of those on which the sun shines is invisible, being obscured by the of the sun. But the Milky Way is the peculiar light of those stars which are shaded by the earth from the sun's rays.

This, too, is obviously impossible. The Milky Way is always unchanged and among the same constellations (for it is clearly a greatest circle), whereas, since the sun does not remain in the same place, what is hidden from it differs at different times. Consequently with the change of the sun's position the Milky Way ought to change its position too: but we find that this does not happen. Besides, if astronomical demonstrations are correct and the size of the sun is greater than that of the earth and the distance of the stars from the earth many times greater than that of the sun (just as the sun is further from the earth than the moon), then the cone made by the rays of the sun would terminate at no great distance from the earth, and the shadow of the earth (what we call night) would not reach the stars. On the contrary, the sun shines on all the stars and the earth screens none of them.

(3) There is a third theory about the Milky Way. Some say that it is a reflection of our sight to the sun, just as they say that the comet is.

But this too is impossible. For if the eye and the mirror and the whole of the object were severally at rest, then the same part of the image would appear at the same point in the mirror. But if the mirror and the object move, keeping the same distance from the eye which is at rest, but at different rates of speed and so not always at the same interval from one another, then it is impossible for the same image always to appear in the same part of the mirror. Now the constellations included in the circle of the Milky Way move; and so does the sun, the object to which our sight is reflected; but we stand still. And the distance of those two from us is constant and uniform, but their distance from one another varies. For the Dolphin sometimes rises at midnight, sometimes in the morning. But in each case the same parts of the Milky Way are found near it. But if it were a reflection and not a genuine affection of these this ought not to be the case.

Again, we can see the Milky Way reflected at night in water and similar mirrors. But under these circumstances it is impossible for our sight to be reflected to the sun.

These considerations show that the Milky Way is not the path of one of the planets, nor the light of imperceptible stars, nor a reflection. And those are the chief theories handed down by others hitherto.

Let us recall our fundamental principle and then explain our views. We have already laid down that the outermost part of what is called the air is potentially fire and that therefore when the air is dissolved by motion, there is separated off a kind of matter-and of this matter we assert that comets consist. We must suppose that what happens is the same as in the case of the comets when the matter does not form independently but is formed by one of the fixed stars or the planets. Then these stars appear to be fringed, because matter of this kind follows their course. In the same way, a certain kind of matter follows the sun, and we explain the halo as a reflection from it when the air is of the right constitution. Now we must assume that what happens in the case of the stars severally happens in the case of the whole of the heavens and all the upper motion. For it is natural to suppose that, if the motion of a single star excites a flame, that of all the stars should have a similar result, and especially in that region in which the stars are biggest and most numerous and nearest to one another. Now the circle of the zodiac dissolves this kind of matter because of the motion of the sun and the planets, and for this reason most comets are found outside the tropic circles. Again, no fringe appears round the sun or moon: for they dissolve such matter too quickly to admit of its formation. But this circle in which the Milky Way appears to our sight is the greatest circle, and its position is such that it extends far outside the tropic circles. Besides the region is full of the biggest and brightest constellations and also of what called 'scattered' stars (you have only to look to see this clearly). So for these reasons all this matter is continually and ceaselessly collecting there. A proof of the theory is this: In the circle itself the light is stronger in that half where the Milky Way is divided, and in it the constellations are more numerous and closer to one another than in the other half; which shows that the cause of the light is the motion of the constellations and nothing else. For if it is found in the circle in which there are most constellations and at that point in the circle at which they are densest and contain the biggest and the most stars, it is natural to suppose that they are the true cause of the affection in question. The circle and the constellations in it may be seen in the diagram. The so-called 'scattered' stars it is not possible to set down in the same way on the sphere because none of them have an evident permanent position; but if you look up to the sky the point is clear. For in this circle alone are the intervals full of these stars: in the other circles there are obvious gaps. Hence if we accept the cause assigned for the appearance of comets as plausible we must assume that the same kind of thing holds good of the Milky Way. For the fringe which in the former case is an affection of a single star here forms in the same way in relation to a whole circle. So if we are to define the Milky Way we may call it 'a fringe attaching to the greatest circle, and due to the matter secreted'. This, as we said before, explains why there are few comets and why they appear rarely; it is because at each revolution of the heavens this matter has always been and is always being separated off and gathered into this region.

We have now explained the phenomena that occur in that part of the terrestrial world which is continuous with the motions of the heavens, namely, shooting-stars and the burning flame, comets and the milky way, these being the chief affections that appear in that region.

Part 9

Let us go on to treat of the region which follows next in order after this and which immediately surrounds the earth. It is the region common to water and air, and the processes attending the formation of water above take place in it. We must consider the principles and causes of all these phenomena too as before. The efficient and chief and first cause is the circle in which the sun moves. For the sun as it approaches or recedes, obviously causes dissipation and condensation and so gives rise to generation and destruction. Now the earth remains but the moisture surrounding it is made to evaporate by the sun's rays and the other heat from above, and rises. But when the heat which was raising it leaves it, in part dispersing to the higher region, in part quenched through rising so far into the upper air, then the vapour cools because its heat is gone and because the place is cold, and condenses again and turns from air into water. And after the water has formed it falls down again to the earth.

The exhalation of water is vapour: air condensing into water is cloud. Mist is what is left over when a cloud condenses into water, and is therefore rather a sign of fine weather than of rain; for mist might be called a barren cloud. So we get a circular process that follows the course of the sun. For according as the sun moves to this side or that, the moisture in this process rises or falls. We must think of it as a river flowing up and down in a circle and made up partly of air, partly of water. When the sun is near, the stream of vapour flows upwards; when it recedes, the stream of water flows down: and the order of sequence, at all events, in this process always remains the same. So if 'Oceanus' had some secret meaning in early writers, perhaps they may have meant this river that flows in a circle about the earth.

So the moisture is always raised by the heat and descends to the earth again when it gets cold. These processes and, in some cases, their varieties are distinguished by special names. When the water falls in small drops it is called a drizzle; when the drops are larger it is rain.

Part 10

Some of the vapour that is formed by day does not rise high because the ratio of the fire that is raising it to the water that is being raised is small. When this cools and descends at night it is called dew and hoar-frost. When the vapour is frozen before it has condensed to water again it is hoar-frost; and this appears in winter and is commoner in cold places. It is dew when the vapour has condensed into water and the heat is not so great as to dry up the moisture that has been raised nor the cold sufficient (owing to the warmth of the climate or season) for the vapour itself to freeze. For dew is more commonly found when the season or the place is warm, whereas the opposite, as has been said, is the case with hoar-frost. For obviously vapour is warmer than water, having still the fire that raised it: consequently more cold is needed to freeze it.

Both dew and hoar-frost are found when the sky is clear and there is no wind. For the vapour could not be raised unless the sky were clear, and if a wind were blowing it could not condense.

The fact that hoar-frost is not found on mountains contributes to prove that these phenomena occur because the vapour does not rise high. One reason for this is that it rises from hollow and watery places, so that the heat that is raising it, bearing as it were too heavy a burden cannot lift it to a great height but soon lets it fall again. A second reason is that the motion of the air is more pronounced at a height, and this dissolves a gathering of this kind.

Everywhere, except in Pontus, dew is found with south winds and not with north winds. There the opposite is the case and it is found with north winds and not with south. The reason is the same as that which explains why dew is found in warm weather and not in cold. For the south wind brings warm, and the north, wintry weather. For the north wind is cold and so quenches the heat of the evaporation. But in Pontus the south wind does not bring warmth enough to cause evaporation, whereas the coldness of the north wind concentrates the heat by a sort of recoil, so that there is more evaporation and not less. This is a thing which we can often observe in other places too. Wells, for instance, give off more vapour in a north than in a south wind. Only the north winds quench the heat before any considerable quantity of vapour has gathered, while in a south wind the evaporation is allowed to accumulate.

Water, once formed, does not freeze on the surface of the earth, in the way that it does in the region of the clouds.

Part 11

From the latter there fall three bodies condensed by cold, namely rain, snow, hail. Two of these correspond to the phenomena on the lower level and are due to the same causes, differing from them only in degree and quantity.

Snow and hoar-frost are one and the same thing, and so are rain and dew: only there is a great deal of the former and little of the latter. For rain is due to the cooling of a great amount of vapour, for the region from which and the time during which the vapour is collected are considerable. But of dew there is little: for the vapour collects for it in a single day and from a small area, as its quick formation and scanty quantity show.

The relation of hoar-frost and snow is the same: when cloud freezes there is snow, when vapour freezes there is hoar-frost. Hence snow is a sign of a cold season or country. For a great deal of heat is still present and unless the cold were overpowering it the cloud would not freeze. For there still survives in it a great deal of the heat which caused the moisture to rise as vapour from the earth.

Hail on the other hand is found in the upper region, but the corresponding phenomenon in the vaporous region near the earth is lacking. For, as we said, to snow in the upper region corresponds hoar-frost in the lower, and to rain in the upper region, dew in the lower. But there is nothing here to correspond to hail in the upper region. Why this is so will be clear when we have explained the nature of hail.

Part 12

But we must go on to collect the facts bearing on the origin of it, both those which raise no difficulties and those which seem paradoxical.

Hail is ice, and water freezes in winter; yet hailstorms occur chiefly in spring and autumn and less often in the late summer, but rarely in winter and then only when the cold is less intense. And in general hailstorms occur in warmer, and snow in colder places. Again, there is a difficulty about water freezing in the upper region. It cannot have frozen before becoming water: and water cannot remain suspended in the air for any space of time. Nor can we say that the case is like that of particles of moisture which are carried up owing to their small size and rest on the iar (the water swimming on the air just as small particles of earth and gold often swim on water). In that case large drops are formed by the union of many small, and so fall down. This cannot take place in the case of hail, since solid bodies cannot coalesce like liquid ones. Clearly then drops of that size were suspended in the air or else they could not have been so large when frozen.

Some think that the cause and origin of hail is this. The cloud is thrust up into the upper atmosphere, which is colder because the reflection of the sun's rays from the earth ceases there, and upon its arrival there the water freezes. They think that this explains why hailstorms are commoner in summer and in warm countries; the heat is greater and it thrusts the clouds further up from the earth. But the fact is that hail does not occur at all at a great height: yet it ought to do so, on their theory, just as we see that snow falls most on high mountains. Again clouds have often been observed moving with a great noise close to the earth, terrifying those who heard and saw them as portents of some catastrophe. Sometimes, too, when such clouds have been seen, without any noise, there follows a violent hailstorm, and the stones are of incredible size, and angular in shape. This shows that they have not been falling for long and that they were frozen near to the earth, and not as that theory would have it. Moreover, where the hailstones are large, the cause of their freezing must be present in the highest degree: for hail is ice as everyone can see. Now those hailstones are large which are angular in shape. And this shows that they froze close to the earth, for those that fall far are worn away by the length of their fall and become round and smaller in size.

It clearly follows that the congelation does not take place because the cloud is thrust up into the cold upper region.

Now we see that warm and cold react upon one another by recoil. Hence in warm weather the lower parts of the earth are cold and in a frost they are warm. The same thing, we must suppose, happens in the air, so that in the warmer seasons the cold is concentrated by the surrounding heat and causes the cloud to go over into water suddenly. (For this reason rain-drops are much larger on warm days than in winter, and showers more violent. A shower is said to be more violent in proportion as the water comes down in a body, and this happens when the condensation takes place quickly,-though this is just the opposite of what Anaxagoras says. He says that this happens when the cloud has risen into the cold air; whereas we say that it happens when the cloud has descended into the warm air, and that the more the further the cloud has descended). But when the cold has been concentrated within still more by the outer heat, it freezes the water it has formed and there is hail. We get hail when the process of freezing is quicker than the descent of the water. For if the water falls in a certain time and the cold is sufficient to freeze it in less, there is no difficulty about its having frozen in the air, provided that the freezing takes place in a shorter time than its fall. The nearer to the earth, and the more suddenly, this process takes place, the more violent is the rain that results and the larger the raindrops and the hailstones because of the shortness of their fall. For the same reason large raindrops do not fall thickly. Hail is rarer in summer than in spring and autumn, though commoner than in winter, because the air is drier in summer, whereas in spring it is still moist, and in autumn it is beginning to grow moist. It is for the same reason that hailstorms sometimes occur in the late summer as we have said.

The fact that the water has previously been warmed contributes to its freezing quickly: for so it cools sooner. Hence many people, when they want to cool hot water quickly, begin by putting it in the sun. So the inhabitants of Pontus when they encamp on the ice to fish (they cut a hole in the ice and then fish) pour warm water round their reeds that it may freeze the quicker, for they use the ice like lead to fix the reeds. Now it is in hot countries and seasons that the water which forms soon grows warm.

It is for the same reason that rain falls in summer and not in winter in Arabia and Ethiopia too, and that in torrents and repeatedly on the same day. For the concentration or recoil due to the extreme heat of the country cools the clouds quickly.

So much for an account of the nature and causes of rain, dew, snow, hoar-frost, and hail.

Part 13

Let us explain the nature of winds, and all windy vapours, also of rivers and of the sea. But here, too, we must first discuss the difficulties involved: for, as in other matters, so in this no theory has been handed down to us that the most ordinary man could not have thought of.

Some say that what is called air, when it is in motion and flows, is wind, and that this same air when it condenses again becomes cloud and water, implying that the nature of wind and water is the same. So they define wind as a motion of the air. Hence some, wishing to say a clever thing, assert that all the winds are one wind, because the air that moves is in fact all of it one and the same; they maintain that the winds appear to differ owing to the region from which the air may happen to flow on each occasion, but really do not differ at all. This is just like thinking that all rivers are one and the same river, and the ordinary unscientific view is better than a scientific theory like this. If all rivers flow from one source, and the same is true in the case of the winds, there might be some truth in this theory; but if it is no more true in the one case than in the other, this ingenious idea is plainly false. What requires investigation is this: the nature of wind and how it originates, its efficient cause and whence they derive their source; whether one ought to think of the wind as issuing from a sort of vessel and flowing until the vessel is empty, as if let out of a wineskin, or, as painters represent the winds, as drawing their source from themselves.

We find analogous views about the origin of rivers. It is thought that the water is raised by the sun and descends in rain and gathers below the earth and so flows from a great reservoir, all the rivers from one, or each from a different one. No water at all is generated, but the volume of the rivers consists of the water that is gathered into such reservoirs in winter. Hence rivers are always fuller in winter than in summer, and some are perennial, others not. Rivers are perennial where the reservoir is large and so enough water has collected in it to last out and not be used up before the winter rain returns. Where the reservoirs are smaller there is less water in the rivers, and they are dried up and their vessel empty before the fresh rain comes on.

But if any one will picture to himself a reservoir adequate to the water that is continuously flowing day by day, and consider the amount of the water, it is obvious that a receptacle that is to contain all the water that flows in the year would be larger than the earth, or, at any rate, not much smaller.

Though it is evident that many reservoirs of this kind do exist in many parts of the earth, yet it is unreasonable for anyone to refuse to admit that air becomes water in the earth for the same reason as it does above it. If the cold causes the vaporous air to condense into water above the earth we must suppose the cold in the earth to produce this same effect, and recognize that there not only exists in it and flows out of it actually formed water, but that water is continually forming in it too.

Again, even in the case of the water that is not being formed from day to day but exists as such, we must not suppose as some do, that rivers have their source in definite subterranean lakes. On the contrary, just as above the earth small drops form and these join others, till finally the water descends in a body as rain, so too we must suppose that in the earth the water at first trickles together little by little, and that the sources of the rivers drip, as it were, out of the earth and then unite. This is proved by facts. When men construct an aqueduct they collect the water in pipes and trenches, as if the earth in the higher ground were sweating the water out. Hence, too, the head-waters of rivers are found to flow from mountains, and from the greatest mountains there flow the most numerous and greatest rivers. Again, most springs are in the neighbourhood of mountains and of high ground, whereas if we except rivers, water rarely appears in the plains. For mountains and high ground, suspended over the country like a saturated sponge, make the water ooze out and trickle together in minute quantities but in many places. They receive a great deal of water falling as rain (for it makes no difference whether a spongy receptacle is concave and turned up or convex and turned down: in either case it will contain the same volume of matter) and, they also cool the vapour that rises and condense it back into water.

Hence, as we said, we find that the greatest rivers flow from the greatest mountains. This can be seen by looking at itineraries: what is recorded in them consists either of things which the writer has seen himself or of such as he has compiled after inquiry from those who have seen them.

In Asia we find that the most numerous and greatest rivers flow from the mountain called Parnassus, admittedly the greatest of all mountains towards the south-east. When you have crossed it you see the outer ocean, the further limit of which is unknown to the dwellers in our world. Besides other rivers there flow from it the Bactrus, the Choaspes, the Araxes: from the last a branch separates off and flows into lake Maeotis as the Tanais. From it, too, flows the Indus, the volume of whose stream is greatest of all rivers. From the Caucasus flows the Phasis, and very many other great rivers besides. Now the Caucasus is the greatest of the mountains that lie to the northeast, both as regards its extent and its height. A proof of its height is the fact that it can be seen from the so-called 'deeps' and from the entrance to the lake. Again, the sun shines on its peaks for a third part of the night before sunrise and again after sunset. Its extent is proved by the fact that thought contains many inhabitable regions which are occupied by many nations and in which there are said to be great lakes, yet they say that all these regions are visible up to the last peak. From Pyrene (this is a mountain towards the west in Celtice) there flow the Istrus and the Tartessus. The latter flows outside the pillars, while the Istrus flows through all Europe into the Euxine. Most of the remaining rivers flow northwards from the Hercynian mountains, which are the greatest in height and extent about that region. In the extreme north, beyond furthest Scythia, are the mountains called Rhipae. The stories about their size are altogether too fabulous: however, they say that the most and (after the Istrus) the greatest rivers flow from them. So, too, in Libya there flow from the Aethiopian mountains the Aegon and the Nyses; and from the so-called Silver Mountain the two greatest of named rivers, the river called Chremetes that flows into the outer ocean, and the main source of the Nile. Of the rivers in the Greek world, the Achelous flows from Pindus, the Inachus from the same mountain; the Strymon, the Nestus, and the Hebrus all three from Scombrus; many rivers, too, flow from Rhodope.

All other rivers would be found to flow in the same way, but we have mentioned these as examples. Even where rivers flow from marshes, the marshes in almost every case are found to lie below mountains or gradually rising ground.

It is clear then that we must not suppose rivers to originate from definite reservoirs: for the whole earth, we might almost say, would not be sufficient (any more than the region of the clouds would be) if we were to suppose that they were fed by actually existing water only and it were not the case that as some water passed out of existence some more came into existence, but rivers always drew their stream from an existing store. Secondly, the fact that rivers rise at the foot of mountains proves that a place transmits the water it contains by gradual percolation of many drops, little by little, and that this is how the sources of rivers originate. However, there is nothing impossible about the existence of such places containing a quantity of water like lakes: only they cannot be big enough to produce the supposed effect. To think that they are is just as absurd as if one were to suppose that rivers drew all their water from the sources we see (for most rivers do flow from springs). So it is no more reasonable to suppose those lakes to contain the whole volume of water than these springs.

That there exist such chasms and cavities in the earth we are taught by the rivers that are swallowed up. They are found in many parts of the earth: in the Peloponnesus, for instance, there are many such rivers in Arcadia. The reason is that Arcadia is mountainous and there are no channels from its valleys to the sea. So these places get full of water, and this, having no outlet, under the pressure of the water that is added above, finds a way out for itself underground. In Greece this kind of thing happens on quite a small scale, but the lake at the foot of the Caucasus, which the inhabitants of these parts call a sea, is considerable. Many great rivers fall into it and it has no visible outlet but issues below the earth off the land of the Coraxi about the so-called 'deeps of Pontus'. This is a place of unfathomable depth in the sea: at any rate no one has yet been able to find bottom there by sounding. At this spot, about three hundred stadia from land, there comes up sweet water over a large area, not all of it together but in three places. And in Liguria a river equal in size to the Rhodanus is swallowed up and appears again elsewhere: the Rhodanus being a navigable river.

Part 14

The same parts of the earth are not always moist or dry, but they change according as rivers come into existence and dry up. And so the relation of land to sea changes too and a place does not always remain land or sea throughout all time, but where there was dry land there comes to be sea, and where there is now sea, there one day comes to be dry land. But we must suppose these changes to follow some order and cycle. The principle and cause of these changes is that the interior of the earth grows and decays, like the bodies of plants and animals. Only in the case of these latter the process does not go on by parts, but each of them necessarily grows or decays as a whole, whereas it does go on by parts in the case of the earth. Here the causes are cold and heat, which increase and diminish on account of the sun and its course. It is owing to them that the parts of the earth come to have a different character, that some parts remain moist for a certain time, and then dry up and grow old, while other parts in their turn are filled with life and moisture. Now when places become drier the springs necessarily give out, and when this happens the rivers first decrease in size and then finally become dry; and when rivers change and disappear in one part and come into existence correspondingly in another, the sea must needs be affected.

If the sea was once pushed out by rivers and encroached upon the land anywhere, it necessarily leaves that place dry when it recedes; again, if the dry land has encroached on the sea at all by a process of silting set up by the rivers when at their full, the time must come when this place will be flooded again.

But the whole vital process of the earth takes place so gradually and in periods of time which are so immense compared with the length of our life, that these changes are not observed, and before their course can be recorded from beginning to end whole nations perish and are destroyed. Of such destructions the most utter and sudden are due to wars; but pestilence or famine cause them too. Famines, again, are either sudden and severe, or else gradual. In the latter case the disappearance of a nation is not noticed because some leave the country while others remain; and this goes on until the land is unable to maintain any inhabitants at all. So a long period of time is likely to elapse from the first departure to the last, and no one remembers and the lapse of time destroys all record even before the last inhabitants have disappeared. In the same way a nation must be supposed to lose account of the time when it first settled in a land that was changing from a marshy and watery state and becoming dry. Here, too, the change is gradual and lasts a long time and men do not remember who came first, or when, or what the land was like when they came. This has been the case with Egypt. Here it is obvious that the land is continually getting drier and that the whole country is a deposit of the river Nile. But because the neighbouring peoples settled in the land gradually as the marshes dried, the lapse of time has hidden the beginning of the process. However, all the mouths of the Nile, with the single exception of that at Canopus, are obviously artificial and not natural. And Egypt was nothing more than what is called Thebes, as Homer, too, shows, modern though he is in relation to such changes. For Thebes is the place that he mentions; which implies that Memphis did not yet exist, or at any rate was not as important as it is now. That this should be so is natural, since the lower land came to be inhabited later than that which lay higher. For the parts that lie nearer to the place where the river is depositing the silt are necessarily marshy for a longer time since the water always lies most in the newly formed land. But in time this land changes its character, and in its turn enjoys a period of prosperity. For these places dry up and come to be in good condition while the places that were formerly well-tempered someday grow excessively dry and deteriorate. This happened to the land of Argos and Mycenae in Greece. In the time of the Trojan wars the Argive land was marshy and could only support a small population, whereas the land of Mycenae was in good condition (and for this reason Mycenae was the superior). But now the opposite is the case, for the reason we have mentioned: the land of Mycenae has become completely dry and barren, while the Argive land that was formerly barren owing to the water has now become fruitful. Now the same process that has taken place in this small district must be supposed to be going on over whole countries and on a large scale.

Men whose outlook is narrow suppose the cause of such events to be change in the universe, in the sense of a coming to be of the world as a whole. Hence they say that the sea being dried up and is growing less, because this is observed to have happened in more places now than formerly. But this is only partially true. It is true that many places are now dry, that formerly were covered with water. But the opposite is true too: for if they look they will find that there are many places where the sea has invaded the land. But we must not suppose that the cause of this is that the world is in process of becoming. For it is absurd to make the universe to be in process because of small and trifling changes, when the bulk and size of the earth are surely as nothing in comparison with the whole world. Rather we must take the cause of all these changes to be that, just as winter occurs in the seasons of the year, so in determined periods there comes a great winter of a great year and with it excess of rain. But this excess does not always occur in the same place. The deluge in the time of Deucalion, for instance, took place chiefly in the Greek world and in it especially about ancient Hellas, the country about Dodona and the Achelous, a river which has often changed its course. Here the Selli dwelt and those who were formerly called Graeci and now Hellenes. When, therefore, such an excess of rain occurs we must suppose that it suffices for a long time. We have seen that some say that the size of the subterranean cavities is what makes some rivers perennial and others not, whereas we maintain that the size of the mountains is the cause, and their density and coldness; for great, dense, and cold mountains catch and keep and create most water: whereas if the mountains that overhang the sources of rivers are small or porous and stony and clayey, these rivers run dry earlier. We must recognize the same kind of thing in this case too. Where such abundance of rain falls in the great winter it tends to make the moisture of those places almost everlasting. But as time goes on places of the latter type dry up more, while those of the former, moist type, do so less: until at last the beginning of the same cycle returns.

Since there is necessarily some change in the whole world, but not in the way of coming into existence or perishing (for the universe is permanent), it must be, as we say, that the same places are not for ever moist through the presence of sea and rivers, nor for ever dry. And the facts prove this. The whole land of the Egyptians, whom we take to be the most ancient of men, has evidently gradually come into existence and been produced by the river. This is clear from an observation of the country, and the facts about the Red Sea suffice to prove it too. One of their kings tried to make a canal to it (for it would have been of no little advantage to them for the whole region to have become navigable; Sesostris is said to have been the first of the ancient kings to try), but he found that the sea was higher than the land. So he first, and Darius afterwards, stopped making the canal, lest the sea should mix with the river water and spoil it. So it is clear that all this part was once unbroken sea. For the same reason Libya-the country of Ammon-is, strangely enough, lower and hollower than the land to the seaward of it. For it is clear that a barrier of silt was formed and after it lakes and dry land, but in course of time the water that was left behind in the lakes dried up and is now all gone. Again the silting up of the lake Maeotis by the rivers has advanced so much that the limit to the size of the ships which can now sail into it to trade is much lower than it was sixty years ago. Hence it is easy to infer that it, too, like most lakes, was originally produced by the rivers and that it must end by drying up entirely.

Again, this process of silting up causes a continuous current through the Bosporus; and in this case we can directly observe the nature of the process. Whenever the current from the Asiatic shore threw up a sandbank, there first formed a small lake behind it. Later it dried up and a second sandbank formed in front of the first and a second lake. This process went on uniformly and without interruption. Now when this has been repeated often enough, in the course of time the strait must become like a river, and in the end the river itself must dry up.

So it is clear, since there will be no end to time and the world is eternal, that neither the Tanais nor the Nile has always been flowing, but that the region whence they flow was once dry: for their effect may be fulfilled, but time cannot. And this will be equally true of all other rivers. But if rivers come into existence and perish and the same parts of the earth were not always moist, the sea must needs change correspondingly. And if the sea is always advancing in one place and receding in another it is clear that the same parts of the whole earth are not always either sea or land, but that all this changes in course of time.

So we have explained that the same parts of the earth are not always land or sea and why that is so: and also why some rivers are perennial and others not.

Book II

Part 1

LET us explain the nature of the sea and the reason why such a large mass of water is salt and the way in which it originally came to be.

The old writers who invented theogonies say that the sea has springs, for they want earth and sea to have foundations and roots of their own. Presumably they thought that this view was grander and more impressive as implying that our earth was an important part of the universe. For they believed that the whole world had been built up round our earth and for its sake, and that the earth was the most important and primary part of it. Others, wiser in human knowledge, give an account of its origin. At first, they say, the earth was surrounded by moisture. Then the sun began to dry it up, part of it evaporated and is the cause of winds and the turnings back of the sun and the moon, while the remainder forms the sea. So the sea is being dried up and is growing less, and will end by being someday entirely dried up. Others say that the sea is a kind of sweat exuded by the earth when the sun heats it, and that this explains its saltiness: for all sweat is salt. Others say that the saltiness is due to the earth. Just as water strained through ashes becomes salt, so the sea owes its saltiness to the admixture of earth with similar properties.

We must now consider the facts which prove that the sea cannot possibly have springs. The waters we find on the earth either flow or are stationary. All flowing water has springs. (By a spring, as we have explained above, we must not understand a source from which waters are ladled as it were from a vessel, but a first point at which the water which is continually forming and percolating gathers.) Stationary water is either that which has collected and has been left standing, marshy pools, for instance, and lakes, which differ merely in size, or else it comes from springs. In this case it is always artificial, I mean as in the case of wells, otherwise the spring would have to be above the outlet. Hence the water from fountains and rivers flows of itself, whereas wells need to be worked artificially. All the waters that exist belong to one or other of these classes.

On the basis of this division we can see that the sea cannot have springs. For it falls under neither of the two classes; it does not flow and it is not artificial; whereas all water from springs must belong to one or other of them. Natural standing water from springs is never found on such a large scale.

Again, there are several seas that have no communication with one another at all. The Red Sea, for instance, communicates but slightly with the ocean outside the straits, and the Hyrcanian and Caspian seas are distinct from this ocean and people dwell all round them. Hence, if these seas had had any springs anywhere they must have been discovered.

It is true that in straits, where the land on either side contracts an open sea into a small space, the sea appears to flow. But this is because it is swinging to and fro. In the open sea this motion is not observed, but where the land narrows and contracts the sea the motion that was imperceptible in the open necessarily strikes the attention.

The whole of the Mediterranean does actually flow. The direction of this flow is determined by the depth of the basins and by the number of rivers. Maeotis flows into Pontus and Pontus into the Aegean. After that the flow of the remaining seas is not so easy to observe. The current of Maeotis and Pontus is due to the number of rivers (more rivers flow into the Euxine and Maeotis than into the whole Mediterranean with its much larger basin), and to their own shallowness. For we find the sea getting deeper and deeper. Pontus is deeper than Maeotis, the Aegean than Pontus, the Sicilian sea than the Aegean; the Sardinian and Tyrrhenic being the deepest of all. (Outside the pillars of Heracles the sea is shallow owing to the mud, but calm, for it lies in a hollow.) We see, then, that just as single rivers flow from mountains, so it is with the earth as a whole: the greatest volume of water flows from the higher regions in the north. Their alluvium makes the northern seas shallow, while the outer seas are deeper. Some further evidence of the height of the northern regions of the earth is afforded by the view of many of the ancient meteorologists. They believed that the sun did not pass below the earth, but round its northern part, and that it was the height of this which obscured the sun and caused night.

So much to prove that there cannot be sources of the sea and to explain its observed flow.

Part 2

We must now discuss the origin of the sea, if it has an origin, and the cause of its salt and bitter taste.

What made earlier writers consider the sea to be the original and main body of water is this. It seems reasonable to suppose that to be the case on the analogy of the other elements. Each of them has a main bulk which by reason of its mass is the origin of that element, and any parts which change and mix with the other elements come from it. Thus the main body of fire is in the upper region; that of air occupies the place next inside the region of fire; while the mass of the earth is that round which the rest of the elements are seen to lie. So we must clearly look for something analogous in the case of water. But here we can find no such single mass, as in the case of the other elements, except the sea. River water is not a unity, nor is it stable, but is seen to be in a continuous process of becoming from day to day. It was this difficulty which made people regard the sea as the origin and source of moisture and of all water. And so we find it maintained that rivers not only flow into the sea but originate from it, the salt water becoming sweet by filtration.

But this view involves another difficulty. If this body of water is the origin and source of all water, why is it salt and not sweet? The reason for this, besides answering this question, will ensure our having a right first conception of the nature of the sea.

The earth is surrounded by water, just as that is by the sphere of air, and that again by the sphere called that of fire (which is the outermost both on the common view and on ours). Now the sun, moving as it does, sets up processes of change and becoming and decay, and by its agency the finest and sweetest water is every day carried up and is dissolved into vapour and rises to the upper region, where it is condensed again by the cold and so returns to the earth. This, as we have said before, is the regular course of nature.

Hence all my predecessors who supposed that the sun was nourished by moisture are absurdly mistaken. Some go on to say that the solstices are due to this, the reason being that the same places cannot always supply the sun with nourishment and that without it he must perish. For the fire we are familiar with lives as long as it is fed, and the only food for fire is moisture. As if the moisture that is raised could reach the sun! or this ascent were really like that performed by flame as it comes into being, and to which they supposed the case of the sun to be analogous! Really there is no similarity. A flame is a process of becoming, involving a constant interchange of moist and dry. It cannot be said to be nourished since it scarcely persists as one and the same for a moment. This cannot be true of the sun; for if it were nourished like that, as they say it is, we should obviously not only have a new sun every day, as Heraclitus says, but a new sun every moment. Again, when the sun causes the moisture to rise, this is like fire heating water. So, as the fire is not fed by the water above it, it is absurd to suppose that the sun feeds on that moisture, even if its heat made all the water in the world evaporate. Again, it is absurd, considering the number and size of the stars, that these thinkers should consider the sun only and overlook the question how the rest of the heavenly bodies subsist. Again, they are met by the same difficulty as those who say that at first the earth itself was moist and the world round the earth was warmed by the sun, and so air was generated and the whole firmament grew, and the air caused winds and solstices. The objection is that we always plainly see the water that has been carried up coming down again. Even if the same amount does not come back in a year or in a given country, yet in a certain period all that has been carried up is returned. This implies that the celestial bodies do not feed on it, and that we cannot distinguish between some air which preserves its character once it is generated and some other which is generated but becomes water again and so perishes; on the contrary, all the moisture alike is dissolved and all of it condensed back into water.

The drinkable, sweet water, then, is light and is all of it drawn up: the salt water is heavy and remains behind, but not in its natural place. For this is a question which has been sufficiently discussed (I mean about the natural place that water, like the other elements, must in reason have), and the answer is this. The place which we see the sea filling is not its natural place but that of water. It seems to belong to the sea because the weight of the salt water makes it remain there, while the sweet, drinkable water which is light is carried up. The same thing happens in animal bodies. Here, too, the food when it enters the body is sweet, yet the residuum and dregs of liquid food are found to be bitter and salt. This is because the sweet and drinkable part of it has been drawn away by the natural animal heat and has passed into the flesh and the other parts of the body according to their several natures. Now just as here it would be wrong for any one to refuse to call the belly the place of liquid food because that disappears from it soon, and to call it the place of the residuum because this is seen to remain, so in the case of our present subject. This place, we say, is the place of water. Hence all rivers and all the water that is generated flow into it: for water flows into the deepest place, and the deepest part of the earth is filled by the sea. Only all the light and sweet part of it is quickly carried off by the sun, while the rest remains for the reason we have explained. It is quite natural that some people should have been puzzled by the old question why such a mass of water leaves no trace anywhere (for the sea does not increase though innumerable and vast rivers are flowing into it every day.) But if one considers the matter the solution is easy. The same amount of water does not take as long to dry up when it is spread out as when it is gathered in a body, and indeed the difference is so great that in the one case it might persist the whole day long while in the other it might all disappear in a moment-as for instance if one were to spread out a cup of water over a large table. This is the case with the rivers: all the time they are flowing their water forms a compact mass, but when it arrives at a vast wide place it quickly and imperceptibly evaporates.

But the theory of the Phaedo about rivers and the sea is impossible. There it is said that the earth is pierced by intercommunicating channels and that the original head and source of all waters is what is called Tartarus-a mass of water about the centre, from which all waters, flowing and standing, are derived. This primary and original water is always surging to and fro, and so it causes the rivers to flow on this side of the earth's centre and on that; for it has no fixed seat but is always oscillating about the centre. Its motion up and down is what fills rivers. Many of these form lakes in various places (our sea is an instance of one of these), but all of them come round again in a circle to the original source of their flow, many at the same point, but some at a point opposite to that from which they issued; for instance, if they started from the other side of the earth's centre, they might return from this side of it. They descend only as far as the centre, for after that all motion is upwards. Water gets its tastes and colours from the kind of earth the rivers happened to flow through.

But on this theory rivers do not always flow in the same sense. For since they flow to the centre from which they issue forth they will not be flowing down any more than up, but in whatever direction the surging of Tartarus inclines to. But at this rate we shall get the proverbial rivers flowing upwards, which is impossible. Again, where is the water that is generated and what goes up again as vapour to come from? For this must all of it simply be ignored, since the quantity of water is always the same and all the water that flows out from the original source flows back to it again. This itself is not true, since all rivers are seen to end in the sea except where one flows into another. Not one of them ends in the earth, but even when one is swallowed up it comes to the surface again. And those rivers are large which flow for a long distance through a lowying country, for by their situation and length they cut off the course of many others and swallow them up. This is why the Istrus and the Nile are the greatest of the rivers which flow into our sea. Indeed, so many rivers fall into them that there is disagreement as to the sources of them both. All of which is plainly impossible on the theory, and the more so as it derives the sea from Tartarus.

Enough has been said to prove that this is the natural place of water and not of the sea, and to explain why sweet water is only found in rivers, while salt water is stationary, and to show that the sea is the end rather than the source of water, analogous to the residual matter of all food, and especially liquid food, in animal bodies.

Part 3

We must now explain why the sea is salt, and ask whether it eternally exists as identically the same body, or whether it did not exist at all once and some day will exist no longer, but will dry up as some people think.

Every one admits this, that if the whole world originated the sea did too; for they make them come into being at the same time. It follows that if the universe is eternal the same must be true of the sea. Anyone who thinks like Democritus that the sea is diminishing and will disappear in the end reminds us of Aesop's tales. His story was that Charybdis had twice sucked in the sea: the first time she made the mountains visible; the second time the islands; and when she sucks it in for the last time she will dry it up entirely. Such a tale is appropriate enough to Aesop in a rage with the ferryman, but not to serious inquirers. Whatever made the sea remain at first, whether it was its weight, as some even of those who hold these views say (for it is easy to see the cause here), or some other reason-clearly the same thing must make it persist for ever. They must either deny that the water raised by the sun will return at all, or, if it does, they must admit that the sea persists for ever or as long as this process goes on, and again, that for the same period of time that sweet water must have been carried up beforehand. So the sea will never dry up: for before that can happen the water that has gone up beforehand will return to it: for if you say that this happens once you must admit its recurrence. If you stop the sun's course there is no drying agency. If you let it go on it will draw up the sweet water as we have said whenever it approaches, and let it descend again when it recedes. This notion about the sea is derived from the fact that many places are found to be drier now than they once were. Why this is so we have explained. The phenomenon is due to temporary excess of rain and not to any process of becoming in which the universe or its parts are involved. Some day the opposite will take place and after that the earth will grow dry once again. We must recognize that this process always goes on thus in a cycle, for that is more satisfactory than to suppose a change in the whole world in order to explain these facts. But we have dwelt longer on this point than it deserves.

To return to the saltiness of the sea: those who create the sea once for all, or indeed generate it at all, cannot account for its saltiness. It makes no difference whether the sea is the residue of all the moisture that is about the earth and has been drawn up by the sun, or whether all the flavour existing in the whole mass of sweet water is due to the admixture of a certain kind of earth. Since the total volume of the sea is the same once the water that evaporated has returned, it follows that it must either have been salt at first too, or, if not at first, then not now either. If it was salt from the very beginning, then we want to know why that was so; and why, if salt water was drawn up then, that is not the case now.

Again, if it is maintained that an admixture of earth makes the sea salt (for they say that earth has many flavours and is washed down by the rivers and so makes the sea salt by its admixture), it is strange that rivers should not be salt too. How can the admixture of this earth have such a striking effect in a great quantity of water and not in each river singly? For the sea, differing in nothing from rivers but in being salt, is evidently simply the totality of river water, and the rivers are the vehicle in which that earth is carried to their common destination.

It is equally absurd to suppose that anything has been explained by calling the sea 'the sweat of the earth', like Empedicles. Metaphors are poetical and so that expression of his may satisfy the requirements of a poem, but as a scientific theory it is unsatisfactory. Even in the case of the body it is a question how the sweet liquid drunk becomes salt sweat whether it is merely by the departure of some element in it which is sweetest, or by the admixture of something, as when water is strained through ashes. Actually the saltiness seems to be due to the same cause as in the case of the residual liquid that gathers in the bladder. That, too, becomes bitter and salt though the liquid we drink and that contained in our food is sweet. If then the bitterness is due in these cases (as with the water strained through lye) to the presence of a certain sort of stuff that is carried along by the urine (as indeed we actually find a salt deposit settling in chamber-pots) and is secreted from the flesh in sweat (as if the departing moisture were washing the stuff out of the body), then no doubt the admixture of something earthy with the water is what makes the sea salt.

Now in the body stuff of this kind, viz. the sediment of food, is due to failure to digest: but how there came to be any such thing in the earth requires explanation. Besides, how can the drying and warming of the earth cause the secretion such a great quantity of water; especially as that must be a mere fragment of what is left in the earth? Again, waiving the question of quantity, why does not the earth sweat now when it happens to be in process of drying? If it did so then, it ought to do so now. But it does not: on the contrary, when it is dry it grows moist, but when it is moist it does not secrete anything at all. How then was it possible for the earth at the beginning when it was moist to sweat as it grew dry? Indeed, the theory that maintains that most of the moisture departed and was drawn up by the sun and that what was left over is the sea is more reasonable; but for the earth to sweat when it is moist is impossible.

Since all the attempts to account for the saltiness of the sea seem unsuccessful let us explain it by the help of the principle we have used already.

Since we recognize two kinds of evaporation, one moist, the other dry, it is clear that the latter must be recognized as the source of phenomena like those we are concerned with.

But there is a question which we must discuss first. Does the sea always remain numerically one and consisting of the same parts, or is it, too, one in form and volume while its parts are in continual change, like air and sweet water and fire? All of these are in a constant state of change, but the form and the quantity of each of them are fixed, just as they are in the case of a flowing river or a burning flame. The answer is clear, and there is no doubt that the same account holds good of all these things alike. They differ in that some of them change more rapidly or more slowly than others; and they all are involved in a process of perishing and becoming which yet affects them all in a regular course.

This being so we must go on to try to explain why the sea is salt. There are many facts which make it clear that this taste is due to the admixture of something. First, in animal bodies what is least digested, the residue of liquid food, is salt and bitter, as we said before. All animal excreta are undigested, but especially that which gathers in the bladder (its extreme lightness proves this; for everything that is digested is condensed), and also sweat; in these then is excreted (along with other matter) an identical substance to which this flavour is due. The case of things burnt is analogous. What heat fails to assimilate becomes the excrementary residue in animal bodies, and, in things burnt, ashes. That is why some people say that it was burnt earth that made the sea salt. To say that it was burnt earth is absurd; but to say that it was something like burnt earth is true. We must suppose that just as in the cases we have described, so in the world as a whole, everything that grows and is naturally generated always leaves an undigested residue, like that of things burnt, consisting of this sort of earth. All the earthy stuff in the dry exhalation is of this nature, and it is the dry exhalation which accounts for its great quantity. Now since, as we have said, the moist and the dry evaporations are mixed, some quantity of this stuff must always be included in the clouds and the water that are formed by condensation, and must redescend to the earth in rain. This process must always go on with such regularity as the sublunary world admits of. and it is the answer to the question how the sea comes to be salt.

It also explains why rain that comes from the south, and the first rains of autumn, are brackish. The south is the warmest of winds and it blows from dry and hot regions. Hence it carries little moist vapour and that is why it is hot. (It makes no difference even if this is not its true character and it is originally a cold wind, for it becomes warm on its way by incorporating with itself a great quantity of dry evaporation from the places it passes over.) The north wind, on the other hand, comb ing from moist regions, is full of vapour and therefore cold. It is dry in our part of the world because it drives the clouds away before it, but in the south it is rainy; just as the south is a dry wind in Libya. So the south wind charges the rain that falls with a great quantity of this stuff. Autumn rain is brackish because the heaviest water must fall first; so that that which contains the greatest quantity of this kind of earth descends quickest.

This, too, is why the sea is warm. Everything that has been exposed to fire contains heat potentially, as we see in the case of lye and ashes and the dry and liquid excreta of animals. Indeed those animals which are hottest in the belly have the hottest excreta.

The action of this cause is continually making the sea more salt, but some part of its saltiness is always being drawn up with the sweet water. This is less than the sweet water in the same ratio in which the salt and brackish element in rain is less than the sweet, and so the saltiness of the sea remains constant on the whole. Salt water when it turns into vapour becomes sweet, and the vapour does not form salt water when it condenses again. This I know by experiment. The same thing is true in every case of the kind: wine and all fluids that evaporate and condense back into a liquid state become water. They all are water modified by a certain admixture, the nature of which determines their flavour. But this subject must be considered on another more suitable occasion.

For the present let us say this. The sea is there and some of it is continually being drawn up and becoming sweet; this returns from above with the rain. But it is now different from what it was when it was drawn up, and its weight makes it sink below the sweet water. This process prevents the sea, as it does rivers, from drying up except from local causes (this must happen to sea and rivers alike). On the other hand the parts neither of the earth nor of the sea remain constant but only their whole bulk. For the same thing is true of the earth as of the sea: some of it is carried up and some comes down with the rain, and both that which remains on the surface and that which comes down again change their situations.

There is more evidence to prove that saltiness is due to the admixture of some substance, besides that which we have adduced. Make a vessel of wax and put it in the sea, fastening its mouth in such a way as to prevent any water getting in. Then the water that percolates through the wax sides of the vessel is sweet, the earthy stuff, the admixture of which makes the water salt, being separated off as it were by a filter. It is this stuff which make salt water heavy (it weighs more than fresh water) and thick. The difference in consistency is such that ships with the same cargo very nearly sink in a river when they are quite fit to navigate in the sea. This circumstance has before now caused loss to shippers freighting their ships in a river. That the thicker consistency is due to an admixture of something is proved by the fact that if you make strong brine by the admixture of salt, eggs, even when they are full, float in it. It almost becomes like mud; such a quantity of earthy matter is there in the sea. The same thing is done in salting fish.

Again if, as is fabled, there is a lake in Palestine, such that if you bind a man or beast and throw it in it floats and does not sink, this would bear out what we have said. They say that this lake is so bitter and salt that no fish live in it and that if you soak clothes in it and shake them it cleans them. The following facts all of them support our theory that it is some earthy stuff in the water which makes it salt. In Chaonia there is a spring of brackish water that flows into a neighbouring river which is sweet but contains no fish. The local story is that when Heracles came from Erytheia driving the oxen and gave the inhabitants the choice, they chose salt in preference to fish. They get the salt from the spring. They boil off some of the water and let the rest stand; when it has cooled and the heat and moisture have evaporated together it gives them salt, not in lumps but loose and light like snow. It is weaker than ordinary salt and added freely gives a sweet taste, and it is not as white as salt generally is. Another instance of this is found in Umbria. There is a place there where reeds and rushes grow. They burn some of these, put the ashes into water and boil it off. When a little water is left and has cooled it gives a quantity of salt.

Most salt rivers and springs must once have been hot. Then the original fire in them was extinguished but the earth through which they percolate preserves the character of lye or ashes. Springs and rivers with all kinds of flavours are found in many places. These flavours must in every case be due to the fire that is or was in them, for if you expose earth to different degrees of heat it assumes various kinds and shades of flavour. It becomes full of alum and lye and other things of the kind, and the fresh water percolates through these and changes its character. Sometimes it becomes acid as in Sicania, a part of Sicily. There they get a salt and acid water which they use as vinegar to season some of their dishes. In the neighbourhood of Lyncus, too, there is a spring of acid water, and in Scythia a bitter spring. The water from this makes the whole of the river into which it flows bitter. These differences are explained by a knowledge of the particular mixtures that determine different savours. But these have been explained in another treatise.

We have now given an account of waters and the sea, why they persist, how they change, what their nature is, and have explained most of their natural operations and affections.

Part 4

Let us proceed to the theory of winds. Its basis is a distinction we have already made. We recognize two kinds of evaporation, one moist, the other dry. The former is called vapour: for the other there is no general name but we must call it a sort of smoke, applying to the whole of it a word that is proper to one of its forms. The moist cannot exist without the dry nor the dry without the moist: whenever we speak of either we mean that it predominates. Now when the sun in its circular course approaches, it draws up by its heat the moist evaporation: when it recedes the cold makes the vapour that had been raised condense back into water which falls and is distributed through the earth. (This explains why there is more rain in winter and more by night than by day: though the fact is not recognized because rain by night is more apt to escape observation than by day.) But there is a great quantity of fire and heat in the earth, and the sun not only draws up the moisture that lies on the surface of it, but warms and dries the earth itself. Consequently, since there are two kinds of evaporation, as we have said, one like vapour, the other like smoke, both of them are necessarily generated. That in which moisture predominates is the source of rain, as we explained before, while the dry evaporation is the source and substance of all winds. That things must necessarily take this course is clear from the resulting phenomena themselves, for the evaporation that is to produce them must necessarily differ; and the sun and the warmth in the earth not only can but must produce these evaporations.

Since the two evaporations are specifically distinct, wind and rain obviously differ and their substance is not the same, as those say who maintain that one and the same air when in motion is wind, but when it condenses again is water. Air, as we have explained in an earlier book, is made up of these as constituents. Vapour is moist and cold (for its fluidity is due to its moistness, and because it derives from water it is naturally cold, like water that has not been warmed): whereas the smoky evaporation is hot and dry. Hence each contributes a part, and air is moist and hot. It is absurd that this air that surrounds us should become wind when in motion, whatever be the source of its motion on the contrary the case of winds is like that of rivers. We do not call water that flows anyhow a river, even if there is a great quantity of it, but only if the flow comes from a spring. So too with the winds; a great quantity of air might be moved by the fall of some large object without flowing from any source or spring.

The facts bear out our theory. It is because the evaporation takes place uninterruptedly but differs in degree and quantity that clouds and winds appear in their natural proportion according to the season; and it is because there is now a great excess of the vaporous, now of the dry and smoky exhalation, that some years are rainy and wet, others windy and dry. Sometimes there is much drought or rain, and it prevails over a great and continuous stretch of country. At other times it is local; the surrounding country often getting seasonable or even excessive rains while there is drought in a certain part; or, contrariwise, all the surrounding country gets little or even no rain while a certain part gets rain in abundance. The reason for all this is that while the same affection is generally apt to prevail over a considerable district because adjacent places (unless there is something special to differentiate them) stand in the same relation to the sun, yet on occasion the dry evaporation will prevail in one part and the moist in another, or conversely. Again the reason for this latter is that each evaporation goes over to that of the neighbouring district: for instance, the dry evaporation circulates in its own place while the moist migrates to the next district or is even driven by winds to some distant place: or else the moist evaporation remains and the dry moves away. Just as in the case of the body when the stomach is dry the lower belly is often in the contrary state, and when it is dry the stomach is moist and cold, so it often happens that the evaporations reciprocally take one another's place and interchange.

Further, after rain wind generally rises in those places where the rain fell, and when rain has come on the wind ceases. These are necessary effects of the principles we have explained. After rain the earth is being dried by its own heat and that from above and gives off the evaporation which we saw to be the material cause of. wind. Again, suppose this secretion is present and wind prevails; the heat is continually being thrown off, rising to the upper region, and so the wind ceases; then the fall in temperature makes vapour form and condense into water. Water also forms and cools the dry evaporation when the clouds are driven together and the cold concentrated in them. These are the causes that make wind cease on the advent of rain, and rain fall on the cessation of wind.

The cause of the predominance of winds from the north and from the south is the same. (Most winds, as a matter of fact, are north winds or south winds.) These are the only regions which the sun does not visit: it approaches them and recedes from them, but its course is always over the-west and the east. Hence clouds collect on either side, and when the sun approaches it provokes the moist evaporation, and when it recedes to the opposite side there are storms and rain. So summer and winter are due to the sun's motion to and from the solstices, and water ascends and falls again for the same reason. Now since most rain falls in those regions towards which and from which the sun turns and these are the north and the south, and since most evaporation must take place where there is the greatest rainfall, just as green wood gives most smoke, and since this evaporation is wind, it is natural that the most and most important winds should come from these quarters. (The winds from the north are called Boreae, those from the south Noti.)

The course of winds is oblique: for though the evaporation rises straight up from the earth, they blow round it because all the surrounding air follows the motion of the heavens. Hence the question might be asked whether winds originate from above or from below. The motion comes from above: before we feel the wind blowing the air betrays its presence if there are clouds or a mist, for their motion shows that the wind has begun to blow before it has actually reached us; and this implies that the source of winds is above. But since wind is defined as 'a quantity of dry evaporation from the earth moving round the earth', it is clear that while the origin of the motion is from above, the matter and the generation of wind come from below. The oblique movement of the rising evaporation is caused from above: for the motion of the heavens determines the processes that are at a distance from the earth, and the motion from below is vertical and every cause is more active where it is nearest to the effect; but in its generation and origin wind plainly derives from the earth.

The facts bear out the view that winds are formed by the gradual union of many evaporations just as rivers derive their sources from the water that oozes from the earth. Every wind is weakest in the spot from which it blows; as they proceed and leave their source at a distance they gather strength. Thus the winter in the north is windless and calm: that is, in the north itself; but, the breeze that blows from there so gently as to escape observation becomes a great wind as it passes on.

We have explained the nature and origin of wind, the occurrence of drought and rains, the reason why rain stops wind and wind rises after rain, the prevalence of north and south winds and also why wind moves in the way it does.

Part 5

The sun both checks the formation of winds and stimulates it. When the evaporation is small in amount and faint the sun wastes it and dissipates by its greater heat the lesser heat contained in the evaporation. It also dries up the earth, the source of the evaporation, before the latter has appeared in bulk: just as, when you throw a little fuel into a great fire, it is often burnt up before giving off any smoke. In these ways the sun checks winds and prevents them from rising at all: it checks them by wasting the evaporation, and prevents their rising by drying up the earth quickly. Hence calm is very apt to prevail about the rising of Orion and lasts until the coming of the Etesiae and their 'forerunners'.

Calm is due to two causes. Either cold quenches the evaporation, for instance a sharp frost: or excessive heat wastes it. In the intermediate periods, too, the causes are generally either that the evaporation has not had time to develop or that it has passed away and there is none as yet to replace it.

Both the setting and the rising of Orion are considered to be treacherous and stormy, because they place at a change of season (namely of summer or winter; and because the size of the constellation makes its rise last over many days) and a state of change is always indefinite and therefore liable to disturbance.

The Etesiae blow after the summer solstice and the rising of the dog-star: not at the time when the sun is closest nor when it is distant; and they blow by day and cease at night. The reason is that when the sun is near it dries up the earth before evaporation has taken place, but when it has receded a little its heat and the evaporation are present in the right proportion; so the ice melts and the earth, dried by its own heat and that of the sun, smokes and vapours. They abate at night because the cold pf the nights checks the melting of the ice. What is frozen gives off no evaporation, nor does that which contains no dryness at all: it is only where something dry contains moisture that it gives off evaporation under the influence of heat.

The question is sometimes asked: why do the north winds which we call the Etesiae blow continuously after the summer solstice, when there are no corresponding south winds after the winter solstice? The facts are reasonable enough: for the so-called 'white south winds' do blow at the corresponding season, though they are not equally continuous and so escape observation and give rise to this inquiry. The reason for this is that the north wind I from the arctic regions which are full of water and snow. The sun thaws them and so the Etesiae blow: after rather than at the summer solstice. (For the greatest heat is developed not when the sun is nearest to the north, but when its heat has been felt for a considerable period and it has not yet receded far. The 'bird winds' blow in the same way after the winter solstice. They, too, are weak Etesiae, but they blow less and later than the Etesiae. They begin to blow only on the seventieth day because the sun is distant and therefore weaker. They do not blow so continuously because only things on the surface of the earth and offering little resistance evaporate then, the thoroughly frozen parts requiring greater heat to melt them. So they blow intermittently till the true Etesiae come on again at the summer solstice: for from that time onwards the wind tends to blow continuously.) But the south wind blows from the tropic of Cancer and not from the antarctic region.

There are two inhabitable sections of the earth: one near our upper, or northern pole, the other near the other or southern pole; and their shape is like that of a tambourine. If you draw lines from the centre of the earth they cut out a drum-shaped figure. The lines form two cones; the base of the one is the tropic, of the other the ever visible circle, their vertex is at the centre of the earth. Two other cones towards the south pole give corresponding segments of the earth. These sections alone are habitable. Beyond the tropics no one can live: for there the shade would not fall to the north, whereas the earth is known to be uninhabitable before the sun is in the zenith or the shade is thrown to the south: and the regions below the Bear are uninhabitable because of the cold.

(The Crown, too, moves over this region: for it is in the zenith when it is on our meridian.)

So we see that the way in which they now describe the geography of the earth is ridiculous. They depict the inhabited earth as round, but both ascertained facts and general considerations show this to be impossible. If we reflect we see that the inhabited region is limited in breadth, while the climate admits of its extending all round the earth. For we meet with no excessive heat or cold in the direction of its length but only in that of its breadth; so that there is nothing to prevent our travelling round the earth unless the extent of the sea presents an obstacle anywhere. The records of journeys by sea and land bear this out. They make the length far greater than the breadth. If we compute these voyages and journeys the distance from the Pillars of Heracles to India exceeds that from Aethiopia to Maeotis and the northernmost Scythians by a ratio of more than 5 to 3, as far as such matters admit of accurate statement. Yet we know the whole breadth of the region we dwell in up to the uninhabited parts: in one direction no one lives because of the cold, in the other because of the heat.

But it is the sea which divides as it seems the parts beyond India from those beyond the Pillars of Heracles and prevents the earth from being inhabited all round.

Now since there must be a region bearing the same relation to the southern pole as the place we live in bears to our pole, it will clearly correspond in the ordering of its winds as well as in other things. So just as we have a north wind here, they must have a corresponding wind from the antarctic. This wind cannot reach us since our own north wind is like a land breeze and does not even reach the limits of the region we live in. The prevalence of north winds here is due to our lying near the north. Yet even here they give out and fail to penetrate far: in the southern sea beyond Libya east and west winds are always blowing alternately, like north and south winds with us. So it is clear that the south wind is not the wind that blows from the south pole. It is neither that nor the wind from the winter tropic. For symmetry would require another wind blowing from the summer tropic, which there is not, since we know that only one wind blows from that quarter. So the south wind clearly blows from the torrid region. Now the sun is so near to that region that it has no water, or snow which might melt and cause Etesiae. But because that place is far more extensive and open the south wind is greater and stronger and warmer than the north and penetrates farther to the north than the north wind does to the south.

The origin of these winds and their relation to one another has now been explained.

Part 6

Let us now explain the position of the winds, their oppositions, which can blow simultaneously with which, and which cannot, their names and number, and any other of their affections that have not been treated in the 'particular questions'. What we say about their position must be followed with the help of the figure. For clearness' sake we have drawn the circle of the horizon, which is round, but it represents the zone in which we live; for that can be divided in the same way. Let us also begin by laying down that those things are locally contrary which are locally most distant from one another, just as things specifically most remote from one another are specific contraries. Now things that face one another from opposite ends of a diameter are locally most distant from one another. (See diagram.)

Let A be the point where the sun sets at the equinox and B, the point opposite, the place where it rises at the equinox. Let there be another diameter cutting this at right angles, and let the point H on it be the north and its diametrical opposite O the south. Let Z be the rising of the sun at the summer solstice and E its setting at the summer solstice; D its rising at the winter solstice, and G its setting at the winter solstice. Draw a diameter from Z to G from D to E. Then since those things are locally contrary which are most distant from one another in space, and points diametrically opposite are most distant from one another, those winds must necessarily be contrary to one another that blow from opposite ends of a diameter.

The names of the winds according to their position are these. Zephyrus is the wind that blows from A, this being the point where the sun sets at the equinox. Its contrary is Apeliotes blowing from B the point where the sun rises at the equinox. The wind blowing from H, the north, is the true north wind, called Aparctias: while Notus blowing from O is its contrary; for this point is the south and O is contrary to H, being diametrically opposite to it. Caecias blows from Z, where the sun rises at the summer solstice. Its contrary is not the wind blowing from E but Lips blowing from G. For Lips blows from the point where the sun sets at the winter solstice and is diametrically opposite to Caecias: so it is its contrary. Eurus blows from D, coming from the point where the sun rises at the winter solstice. It borders on Notus, and so we often find that people speak of 'Euro-Noti'. Its contrary is not Lips blowing from G but the wind that blows from E which some call Argestes, some Olympias, and some Sciron. This blows from the point where the sun sets at the summer solstice, and is the only wind that is diametrically opposite to Eurus. These are the winds that are diametrically opposite to one another and their contraries.

There are other winds which have no contraries. The wind they call Thrascias, which lies between Argestes and Aparctias, blows from I; and the wind called Meses, which lies between Caecias and Aparctias, from K. (The line IK nearly coincides with the ever visible circle, but not quite.) These winds have no contraries. Meses has not, or else there would be a wind blowing from the point M which is diametrically opposite. Thrascias corresponding to the point I has not, for then there would be a wind blowing from N, the point which is diametrically opposite. (But perhaps a local wind which the inhabitants of those parts call Phoenicias blows from that point.)

These are the most important and definite winds and these their places.

There are more winds from the north than from the south. The reason for this is that the region in which we live lies nearer to the north. Also, much more water and snow is pushed aside into this quarter because the other lies under the sun and its course. When this thaws and soaks into the earth and is exposed to the heat of the sun and the earth it necessarily causes evaporation to rise in greater quantities and over a greater space.

Of the winds we have described Aparctias is the north wind in the strict sense. Thrascias and Meses are north winds too. (Caecias is half north and half east.) South are that which blows from due south and Lips. East, the wind from the rising of the sun at the equinox and Eurus. Phoenicias is half south and half east. West, the wind from the true west and that called Argestes.

More generally these winds are classified as northerly or southerly. The west winds are counted as northerly, for they blow from the place of sunset and are therefore colder; the east winds as southerly, for they are warmer because they blow from the place of sunrise. So the distinction of cold and hot or warm is the basis for the division of the winds into northerly and southerly. East winds are warmer than west winds because the sun shines on the east longer, whereas it leaves the west sooner and reaches it later.

Since this is the distribution of the winds it is clear that contrary winds cannot blow simultaneously. They are diametrically opposite to one another and one of the two must be overpowered and cease. Winds that are not diametrically opposite to one another may blow simultaneously: for instance the winds from Z and from D. Hence it sometimes happens that both of them, though different winds and blowing from different quarters, are favourable to sailors making for the same point.

Contrary winds commonly blow at opposite seasons. Thus Caecias and in general the winds north of the summer solstice blow about the time of the spring equinox, but about the autumn equinox Lips; and Zephyrus about the summer solstice, but about the winter solstice Eurus.

Aparctias, Thrascias, and Argestes are the winds that fall on others most and stop them. Their source is so close to us that they are greater and stronger than other winds. They bring fair weather most of all winds for the same reason, for, blowing as they do, from close at hand, they overpower the other winds and stop them; they also blow away the clouds that are forming and leave a clear sky-unless they happen to be very cold. Then they do not bring fair weather, but being colder than they are strong they condense the clouds before driving them away.

Caecias does not bring fair weather because it returns upon itself. Hence the saying: 'Bringing it on himself as Caecias does clouds.'

When they cease, winds are succeeded by their neighbours in the direction of the movement of the sun. For an effect is most apt to be produced in the neighbourhood of its cause, and the cause of winds moves with the sun.

Contrary winds have either the same or contrary effects. Thus Lips and Caecias, sometimes called Hellespontias, are both rainy gestes and Eurus are dry: the latter being dry at first and rainy afterwards. Meses and Aparctias are coldest and bring most snow. Aparctias, Thrascias, and Argestes bring hail. Notus, Zephyrus, and Eurus are hot. Caecias covers the sky with heavy clouds, Lips with lighter ones. Caecias does this because it returns upon itself and combines the qualities of Boreas and Eurus. By being cold it condenses and gathers the vaporous air, and because it is easterly it carries with it and drives before it a great quantity of such matter. Aparctias, Thrascias, and Argestes bring fair weather for the reason we have explained before. These winds and Meses are most commonly accompanied by lightning. They are cold because they blow from the north, and lightning is due to cold, being ejected when the clouds contract. Some of these same bring hail with them for the same reason; namely, that they cause a sudden condensation.

Hurricanes are commonest in autumn, and next in spring: Aparctias, Thrascias, and Argestes give rise to them most. This is because hurricanes are generally formed when some winds are blowing and others fall on them; and these are the winds which are most apt to fall on others that are blowing; the reason for which, too, we have explained before.

The Etesiae veer round: they begin from the north, and become for dwellers in the west Thrasciae, Argestae, and Zephyrus (for Zephyrus belongs to the north). For dwellers in the east they veer round as far as Apeliotes.

So much for the winds, their origin and nature and the properties common to them all or peculiar to each.

Part 7

We must go on to discuss earthquakes next, for their cause is akin to our last subject.

The theories that have been put forward up to the present date are three, and their authors three men, Anaxagoras of Clazomenae, and before him Anaximenes of Miletus, and later Democritus of Abdera.

Anaxagoras says that the ether, which naturally moves upwards, is caught in hollows below the earth and so shakes it, for though the earth is really all of it equally porous, its surface is clogged up by rain. This implies that part of the whole sphere is 'above' and part 'below': 'above' being the part on which we live, 'below' the other.

This theory is perhaps too primitive to require refutation. It is absurd to think of up and down otherwise than as meaning that heavy bodies move to the earth from every quarter, and light ones, such as fire, away from it; especially as we see that, as far as our knowledge of the earth goes, the horizon always changes with a change in our position, which proves that the earth is convex and spherical. It is absurd, too, to maintain that the earth rests on the air because of its size, and then to say that impact upwards from below shakes it right through. Besides he gives no account of the circumstances attendant on earthquakes: for not every country or every season is subject to them.

Democritus says that the earth is full of water and that when a quantity of rain-water is added to this an earthquake is the result. The hollows in the earth being unable to admit the excess of water it forces its way in and so causes an earthquake. Or again, the earth as it dries draws the water from the fuller to the emptier parts, and the inrush of the water as it changes its place causes the earthquake.

Anaximenes says that the earth breaks up when it grows wet or dry, and earthquakes are due to the fall of these masses as they break away. Hence earthquakes take place in times of drought and again of heavy rain, since, as we have explained, the earth grows dry in time of drought and breaks up, whereas the rain makes it sodden and destroys its cohesion.

But if this were the case the earth ought to be found to be sinking in many places. Again, why do earthquakes frequently occur in places which are not excessively subject to drought or rain, as they ought to be on the theory? Besides, on this view, earthquakes ought always to be getting fewer, and should come to an end entirely some day: the notion of contraction by packing together implies this. So this is impossible-the theory must be impossible too.

Part 8

We have already shown that wet and dry must both give rise to an evaporation: earthquakes are a necessary consequence of this fact. The earth is essentially dry, but rain fills it with moisture. Then the sun and its own fire warm it and give rise to a quantity of wind both outside and inside it. This wind sometimes flows outwards in a single body, sometimes inwards, and sometimes it is divided. All these are necessary laws. Next we must find out what body has the greatest motive force. This will certainly be the body that naturally moves farthest and is most violent. Now that which has the most rapid motion is necessarily the most violent; for its swiftness gives its impact the greatest force. Again, the rarest body, that which can most readily pass through every other body, is that which naturally moves farthest. Wind satisfies these conditions in the highest degree (fire only becomes flame and moves rapidly when wind accompanies it): so that not water nor earth is the cause of earthquakes but wind-that is, the inrush of the external evaporation into the earth.

Hence, since the evaporation generally follows in a continuous body in the direction in which it first started, and either all of it flows inwards or all outwards, most earthquakes and the greatest are accompanied by calm. It is true that some take place when a wind is blowing, but this presents no difficulty. We sometimes find several winds blowing simultaneously. If one of these enters the earth we get an earthquake attended by wind. Only these earthquakes are less severe because their source and cause is divided.

Again, most earthquakes and the severest occur at night or, if by day, about noon, that being generally the calmest part of the day. For when the sun exerts its full power (as it does about noon) it shuts the evaporation into the earth. Night, too, is calmer than day. The absence of the sun makes the evaporation return into the earth like a sort of ebb tide, corresponding to the outward flow; especially towards dawn, for the winds, as a rule, begin to blow then, and if their source changes about like the Euripus and flows inwards the quantity of wind in the earth is greater and a more violent earthquake results.

The severest earthquakes take place where the sea is full of currents or the earth spongy and cavernous: so they occur near the Hellespont and in Achaea and Sicily, and those parts of Euboea which correspond to our description-where the sea is supposed to flow in channels below the earth. The hot springs, too, near Aedepsus are due to a cause of this kind. It is the confined character of these places that makes them so liable to earthquakes. A great and therefore violent wind is developed, which would naturally blow away from the earth: but the onrush of the sea in a great mass thrusts it back into the earth. The countries that are spongy below the surface are exposed to earthquakes because they have room for so much wind.

For the same reason earthquakes usually take place in spring and autumn and in times of wet and of drought-because these are the windiest seasons. Summer with its heat and winter with its frost cause calm: winter is too cold, summer too dry for winds to form. In time of drought the air is full of wind; drought is just the predominance of the dry over the moist evaporation. Again, excessive rain causes more of the evaporation to form in the earth. Then this secretion is shut up in a narrow compass and forced into a smaller space by the water that fills the cavities. Thus a great wind is compressed into a smaller space and so gets the upper hand, and then breaks out and beats against the earth and shakes it violently.

We must suppose the action of the wind in the earth to be analogous to the tremors and throbbings caused in us by the force of the wind contained in our bodies. Thus some earthquakes are a sort of tremor, others a sort of throbbing. Again, we must think of an earthquake as something like the tremor that often runs through the body after passing water as the wind returns inwards from without in one volume.

The force wind can have may be gathered not only from what happens in the air (where one might suppose that it owed its power to produce such effects to its volume), but also from what is observed in animal bodies. Tetanus and spasms are motions of wind, and their force is such that the united efforts of many men do not succeed in overcoming the movements of the patients. We must suppose, then (to compare great things with small), that what happens in the earth is just like that. Our theory has been verified by actual observation in many places. It has been known to happen that an earthquake has continued until the wind that caused it burst through the earth into the air and appeared visibly like a hurricane. This happened lately near Heracleia in Pontus and some time past at the island Hiera, one of the group called the Aeolian islands. Here a portion of the earth swelled up and a lump like a mound rose with a noise: finally it burst, and a great wind came out of it and threw up live cinders and ashes which buried the neighbouring town of Lipara and reached some of the towns in Italy. The spot where this eruption occurred is still to be seen.

Indeed, this must be recognized as the cause of the fire that is generated in the earth: the air is first broken up in small particles and then the wind is beaten about and so catches fire.

A phenomenon in these islands affords further evidence of the fact that winds move below the surface of the earth. When a south wind is going to blow there is a premonitory indication: a sound is heard in the places from which the eruptions issue. This is because the sea is being pushed on from a distance and its advance thrusts back into the earth the wind that was issuing from it. The reason why there is a noise and no earthquake is that the underground spaces are so extensive in proportion to the quantity of the air that is being driven on that the wind slips away into the void beyond.

Again, our theory is supported by the facts that the sun appears hazy and is darkened in the absence of clouds, and that there is sometimes calm and sharp frost before earthquakes at sunrise. The sun is necessarily obscured and darkened when the evaporation which dissolves and rarefies the air begins to withdraw into the earth. The calm, too, and the cold towards sunrise and dawn follow from the theory. The calm we have already explained. There must as a rule be calm because the wind flows back into the earth: again, it must be most marked before the more violent earthquakes, for when the wind is not part outside earth, part inside, but moves in a single body, its strength must be greater. The cold comes because the evaporation which is naturally and essentially hot enters the earth. (Wind is not recognized to be hot, because it sets the air in motion, and that is full of a quantity of cold vapour. It is the same with the breath we blow from our mouth: close by it is warm, as it is when we breathe out through the mouth, but there is so little of it that it is scarcely noticed, whereas at a distance it is cold for the same reason as wind.) Well, when this evaporation disappears into the earth the vaporous exhalation concentrates and causes cold in any place in which this disappearance occurs.

A sign which sometimes precedes earthquakes can be explained in the same way. Either by day or a little after sunset, in fine weather, a little, light, long-drawn cloud is seen, like a long very straight line. This is because the wind is leaving the air and dying down. Something analogous to this happens on the sea-shore. When the sea breaks in great waves the marks left on the sand are very thick and crooked, but when the sea is calm they are slight and straight (because the secretion is small). As the sea is to the shore so the wind is to the cloudy air; so, when the wind drops, this very straight and thin cloud is left, a sort of wave-mark in the air.

An earthquake sometimes coincides with an eclipse of the moon for the same reason. When the earth is on the point of being interposed, but the light and heat of the sun has not quite vanished from the air but is dying away, the wind which causes the earthquake before the eclipse, turns off into the earth, and calm ensues. For there often are winds before eclipses: at nightfall if the eclipse is at midnight, and at midnight if the eclipse is at dawn. They are caused by the lessening of the warmth from the moon when its sphere approaches the point at which the eclipse is going to take place. So the influence which restrained and quieted the air weakens and the air moves again and a wind rises, and does so later, the later the eclipse.

A severe earthquake does not stop at once or after a single shock, but first the shocks go on, often for about forty days; after that, for one or even two years it gives premonitory indications in the same place. The severity of the earthquake is determined by the quantity of wind and the shape of the passages through which it flows. Where it is beaten back and cannot easily find its way out the shocks are most violent, and there it must remain in a cramped space like water that cannot escape. Any throbbing in the body does not cease suddenly or quickly, but by degrees according as the affection passes off. So here the agency which created the evaporation and gave it an impulse to motion clearly does not at once exhaust the whole of the material from which it forms the wind which we call an earthquake. So until the rest of this is exhausted the shocks must continue, though more gently, and they must go on until there is too little of the evaporation left to have any perceptible effect on the earth at all.

Subterranean noises, too, are due to the wind; sometimes they portend earthquakes but sometimes they have been heard without any earthquake following. Just as the air gives off various sounds when it is struck, so it does when it strikes other things; for striking involves being struck and so the two cases are the same. The sound precedes the shock because sound is thinner and passes through things more readily than wind. But when the wind is too weak by reason of thinness to cause an earthquake the absence of a shock is due to its filtering through readily, though by striking hard and hollow masses of different shapes it makes various noises, so that the earth sometimes seems to 'bellow' as the portentmongers say.

Water has been known to burst out during an earthquake. But that does not make water the cause of the earthquake. The wind is the efficient cause whether it drives the water along the surface or up from below: just as winds are the causes of waves and not waves of winds. Else we might as well say that earth was the cause; for it is upset in an earthquake, just like water (for effusion is a form of upsetting). No, earth and water are material causes (being patients, not agents): the true cause is the wind.

The combination of a tidal wave with an earthquake is due to the presence of contrary winds. It occurs when the wind which is shaking the earth does not entirely succeed in driving off the sea which another wind is bringing on, but pushes it back and heaps it up in a great mass in one place. Given this situation it follows that when this wind gives way the whole body of the sea, driven on by the other wind, will burst out and overwhelm the land. This is what happened in Achaea. There a south wind was blowing, but outside a north wind; then there was a calm and the wind entered the earth, and then the tidal wave came on and simultaneously there was an earthquake. This was the more violent as the sea allowed no exit to the wind that had entered the earth, but shut it in. So in their struggle with one another the wind caused the earthquake, and the wave by its settling down the inundation.

Earthquakes are local and often affect a small district only; whereas winds are not local. Such phenomena are local when the evaporations at a given place are joined by those from the next and unite; this, as we explained, is what happens when there is drought or excessive rain locally. Now earthquakes do come about in this way but winds do not. For earthquakes, rains, and droughts have their source and origin inside the earth, so that the sun is not equally able to direct all the evaporations in one direction. But on the evaporations in the air the sun has more influence so that, when once they have been given an impulse by its motion, which is determined by its various positions, they flow in one direction.

When the wind is present in sufficient quantity there is an earthquake. The shocks are horizontal like a tremor; except occasionally, in a few places, where they act vertically, upwards from below, like a throbbing. It is the vertical direction which makes this kind of earthquake so rare. The motive force does not easily accumulate in great quantity in the position required, since the surface of the earth secretes far more of the evaporation than its depths. Wherever an earthquake of this kind does occur a quantity of stones comes to the surface of the earth (as when you throw up things in a winnowing fan), as we see from Sipylus and the Phlegraean plain and the district in Liguria, which were devastated by this kind of earthquake.

Islands in the middle of the sea are less exposed to earthquakes than those near land. First, the volume of the sea cools the evaporations and overpowers them by its weight and so crushes them. Then, currents and not shocks are produced in the sea by the action of the winds. Again, it is so extensive that evaporations do not collect in it but issue from it, and these draw the evaporations from the earth after them. Islands near the continent really form part of it: the intervening sea is not enough to make any difference; but those in the open sea can only be shaken if the whole of the sea that surrounds them is shaken too.

We have now explained earthquakes, their nature and cause, and the most important of the circumstances attendant on their appearance.

Part 9

Let us go on to explain lightning and thunder, and further whirlwind, fire-wind, and thunderbolts: for the cause of them all is the same.

As we have said, there are two kinds of exhalation, moist and dry, and the atmosphere contains them both potentially. It, as we have said before, condenses into cloud, and the density of the clouds is highest at their upper limit. (For they must be denser and colder on the side where the heat escapes to the upper region and leaves them. This explains why hurricanes and thunderbolts and all analogous phenomena move downwards in spite of the fact that everything hot has a natural tendency upwards. Just as the pips that we squeeze between our fingers are heavy but often jump upwards: so these things are necessarily squeezed out away from the densest part of the cloud.) Now the heat that escapes disperses to the up region. But if any of the dry exhalation is caught in the process as the air cools, it is squeezed out as the clouds contract, and collides in its rapid course with the neighbouring clouds, and the sound of this collision is what we call thunder. This collision is analogous, to compare small with great, to the sound we hear in a flame which men call the laughter or the threat of Hephaestus or of Hestia. This occurs when the wood dries and cracks and the exhalation rushes on the flame in a body. So in the clouds, the exhalation is projected and its impact on dense clouds causes thunder: the variety of the sound is due to the irregularity of the clouds and the hollows that intervene where their density is interrupted. This then, is thunder, and this its cause.

It usually happens that the exhalation that is ejected is inflamed and burns with a thin and faint fire: this is what we call lightning, where we see as it were the exhalation coloured in the act of its ejection. It comes into existence after the collision and the thunder, though we see it earlier because sight is quicker than hearing. The rowing of triremes illustrates this: the oars are going back again before the sound of their striking the water reaches us.

However, there are some who maintain that there is actually fire in the clouds. Empedocles says that it consists of some of the sun's rays which are intercepted: Anaxagoras that it is part of the upper ether (which he calls fire) which has descended from above. Lightning, then, is the gleam of this fire, and thunder the hissing noise of its extinction in the cloud.

But this involves the view that lightning actually is prior to thunder and does not merely appear to be so. Again, this intercepting of the fire is impossible on either theory, but especially it is said to be drawn down from the upper ether. Some reason ought to be given why that which naturally ascends should descend, and why it should not always do so, but only when it is cloudy. When the sky is clear there is no lightning: to say that there is, is altogether wanton.

The view that the heat of the sun's rays intercepted in the clouds is the cause of these phenomena is equally unattractive: this, too, is a most careless explanation. Thunder, lightning, and the rest must have a separate and determinate cause assigned to them on which they ensue. But this theory does nothing of the sort. It is like supposing that water, snow, and hail existed all along and were produced when the time came and not generated at all, as if the atmosphere brought each to hand out of its stock from time to time. They are concretions in the same way as thunder and lightning are discretions, so that if it is true of either that they are not generated but pre-exist, the same must be true of the other. Again, how can any distinction be made about the intercepting between this case and that of interception in denser substances such as water? Water, too, is heated by the sun and by fire: yet when it contracts again and grows cold and freezes no such ejection as they describe occurs, though it ought on their the. to take place on a proportionate scale. Boiling is due to the exhalation generated by fire: but it is impossible for it to exist in the water beforehand; and besides they call the noise 'hissing', not 'boiling'. But hissing is really boiling on a small scale: for when that which is brought into contact with moisture and is in process of being extinguished gets the better of it, then it boils and makes the noise in question. Some-Cleidemus is one of them-say that lightning is nothing objective but merely an appearance. They compare it to what happens when you strike the sea with a rod by night and the water is seen to shine. They say that the moisture in the cloud is beaten about in the same way, and that lightning is the appearance of brightness that ensues.

This theory is due to ignorance of the theory of reflection, which is the real cause of that phenomenon. The water appears to shine when struck because our sight is reflected from it to some bright object: hence the phenomenon occurs mainly by night: the appearance is not seen by day because the daylight is too in, tense and obscures it.

These are the theories of others about thunder and lightning: some maintaining that lightning is a reflection, the others that lightning is fire shining through the cloud and thunder its extinction, the fire not being generated in each case but existing beforehand. We say that the same stuff is wind on the earth, and earthquake under it, and in the clouds thunder. The essential constituent of all these phenomena is the same: namely, the dry exhalation. If it flows in one direction it is wind, in another it causes earthquakes; in the clouds, when they are in a process of change and contract and condense into water, it is ejected and causes thunder and lightning and the other phenomena of the same nature.

So much for thunder and lightning.

Book III

Part 1

LET us explain the remaining operations of this secretion in the same way as we have treated the rest. When this exhalation is secreted in small and scattered quantities and frequently, and is transitory, and its constitution rare, it gives rise to thunder and lightning. But if it is secreted in a body and is denser, that is, less rare, we get a hurricane. The fact that it issues in body explains its violence: it is due to the rapidity of the secretion. Now when this secretion issues in a great and continuous current the result corresponds to what we get when the opposite development takes place and rain and a quantity of water are produced. As far as the matter from which they are developed goes both sets of phenomena are the same. As soon as a stimulus to the development of either potentiality appears, that of which there is the greater quantity present in the cloud is at once secreted from it, and there results either rain, or, if the other exhalation prevails, a hurricane.

Sometimes the exhalation in the cloud, when it is being secreted, collides with another under circumstances like those found when a wind is forced from an open into a narrow space in a gateway or a road. It often happens in such cases that the first part of the moving body is deflected because of the resistance due either to the narrowness or to a contrary current, and so the wind forms a circle and eddy. It is prevented from advancing in a straight line: at the same time it is pushed on from behind; so it is compelled to move sideways in the direction of least resistance. The same thing happens to the next part, and the next, and so on, till the series becomes one, that is, till a circle is formed: for if a figure is described by a single motion that figure must itself be one. This is how eddies are generated on the earth, and the case is the same in the clouds as far as the beginning of them goes. Only here (as in the case of the hurricane which shakes off the cloud without cessation and becomes a continuous wind) the cloud follows the exhalation unbroken, and the exhalation, failing to break away from the cloud because of its density, first moves in a circle for the reason given and then descends, because clouds are always densest on the side where the heat escapes. This phenomenon is called a whirlwind when it is colourless; and it is a sort of undigested hurricane. There is never a whirlwind when the weather is northerly, nor a hurricane when there is snow. The reason is that all these phenomena are 'wind', and wind is a dry and warm evaporation. Now frost and cold prevail over this principle and quench it at its birth: that they do prevail is clear or there could be no snow or northerly rain, since these occur when the cold does prevail.

So the whirlwind originates in the failure of an incipient hurricane to escape from its cloud: it is due to the resistance which generates the eddy, and it consists in the spiral which descends to the earth and drags with it the cloud which it cannot shake off. It moves things by its wind in the direction in which it is blowing in a straight line, and whirls round by its circular motion and forcibly snatches up whatever it meets.

When the cloud burns as it is drawn downwards, that is, when the exhalation becomes rarer, it is called a fire-wind, for its fire colours the neighbouring air and inflames it.

When there is a great quantity of exhalation and it is rare and is squeezed out in the cloud itself we get a thunderbolt. If the exhalation is exceedingly rare this rareness prevents the thunderbolt from scorching and the poets call it 'bright': if the rareness is less it does scorch and they call it 'smoky'. The former moves rapidly because of its rareness, and because of its rapidity passes through an object before setting fire to it or dwelling on it so as to blacken it: the slower one does blacken the object, but passes through it before it can actually burn it. Further, resisting substances are affected, unresisting ones are not. For instance, it has happened that the bronze of a shield has been melted while the woodwork remained intact because its texture was so loose that the exhalation filtered through without affecting it. So it has passed through clothes, too, without burning them, and has merely reduced them to shreds.

Such evidence is enough by itself to show that the exhalation is at work in all these cases, but we sometimes get direct evidence as well, as in the case of the conflagration of the temple at Ephesus which we lately witnessed. There independent sheets of flame left the main fire and were carried bodily in many directions. Now that smoke is exhalation and that smoke burns is certain, and has been stated in another place before; but when the flame moves bodily, then we have ocular proof that smoke is exhalation. On this occasion what is seen in small fires appeared on a much larger scale because of the quantity of matter that was burning. The beams which were the source of the exhalation split, and a quantity of it rushed in a body from the place from which it issued forth and went up in a blaze: so that the flame was actually seen moving through the air away and falling on the houses. For we must recognize that exhalation accompanies and precedes thunderbolts though it is colourless and so invisible. Hence, where the thunderbolt is going to strike, the object moves before it is struck, showing that the exhalation leads the way and falls on the object first. Thunder, too, splits things not by its noise but because the exhalation that strikes the object and that which makes the noise are ejected simultaneously. This exhalation splits the thing it strikes but does not scorch it at all.

We have now explained thunder and lightning and hurricane, and further firewinds, whirlwinds, and thunderbolts, and shown that they are all of them forms of the same thing and wherein they all differ.

Part 2

Let us now explain the nature and cause of halo, rainbow, mock suns, and rods, since the same account applies to them all.

We must first describe the phenomena and the circumstances in which each of them occurs. The halo often appears as a complete circle: it is seen round the sun and the moon and bright stars, by night as well as by day, and at midday or in the afternoon, more rarely about sunrise or sunset.

The rainbow never forms a full circle, nor any segment greater than a semicircle. At sunset and sunrise the circle is smallest and the segment largest: as the sun rises higher the circle is larger and the segment smaller. After the autumn equinox in the shorter days it is seen at every hour of the day, in the summer not about midday. There are never more than two rainbows at one time. Each of them is three-coloured; the colours are the same in both and their number is the same, but in the outer rainbow they are fainter and their position is reversed. In the inner rainbow the first and largest band is red; in the outer rainbow the band that is nearest to this one and smallest is of the same colour: the other bands correspond on the same principle. These are almost the only colours which painters cannot manufacture: for there are colours which they create by mixing, but no mixing will give red, green, or purple. These are the colours of the rainbow, though between the red and the green an orange colour is often seen.

Mock suns and rods are always seen by the side of the sun, not above or below it nor in the opposite quarter of the sky. They are not seen at night but always in the neighbourhood of the sun, either as it is rising or setting but more commonly towards sunset. They have scarcely ever appeared when the sun was on the meridian, though this once happened in Bosporus where two mock suns rose with the sun and followed it all through the day till sunset.

These are the facts about each of these phenomena: the cause of them all is the same, for they are all reflections. But they are different varieties, and are distinguished by the surface from which and the way in which the reflection to the sun or some other bright object takes place.

The rainbow is seen by day, and it was formerly thought that it never appeared by night as a moon rainbow. This opinion was due to the rarity of the occurrence: it was not observed, for though it does happen it does so rarely. The reason is that the colours are not so easy to see in the dark and that many other conditions must coincide, and all that in a single day in the month. For if there is to be one it must be at full moon, and then as the moon is either rising or setting. So we have only met with two instances of a moon rainbow in more than fifty years.

We must accept from the theory of optics the fact that sight is reflected from air and any object with a smooth surface just as it is from water; also that in some mirrors the forms of things are reflected, in others only their colours. Of the latter kind are those mirrors which are so small as to be indivisible for sense. It is impossible that the figure of a thing should be reflected in them, for if it is the mirror will be sensibly divisible since divisibility is involved in the notion of figure. But since something must be reflected in them and figure cannot be, it remains that colour alone should be reflected. The colour of a bright object sometimes appears bright in the reflection, but it sometimes, either owing to the admixture of the colour of the mirror or to weakness of sight, gives rise to the appearance of another colour.

However, we must accept the account we have given of these things in the theory of sensation, and take some things for granted while we explain others.

Part 3

Let us begin by explaining the shape of the halo; why it is a circle and why it appears round the sun or the moon or one of the other stars: the explanation being in all these cases the same.

Sight is reflected in this way when air and vapour are condensed into a cloud and the condensed matter is uniform and consists of small parts. Hence in itself it is a sign of rain, but if it fades away, of fine weather, if it is broken up, of wind. For if it does not fade away and is not broken up but is allowed to attain its normal state, it is naturally a sign of rain since it shows that a process of condensation is proceeding which must, when it is carried to an end, result in rain. For the same reason these haloes are the darkest. It is a sign of wind when it is broken up because its breaking up is due to a wind which exists there but has not reached us. This view finds support in the fact that the wind blows from the quarter in which the main division appears in the halo. Its fading away is a sign of fine weather because if the air is not yet in a state to get the better of the heat it contains and proceed to condense into water, this shows that the moist vapour has not yet separated from the dry and firelike exhalation: and this is the cause of fine weather.

So much for the atmospheric conditions under which the reflection takes place. The reflection is from the mist that forms round the sun or the moon, and that is why the halo is not seen opposite the sun like the rainbow.

Since the reflection takes place in the same way from every point the result is necessarily a circle or a segment of a circle: for if the lines start from the same point and end at the same point and are equal, the points where they form an angle will always lie on a circle.

Let AGB and AZB and ADB be lines each of which goes from the point A to the point B and forms an angle. Let the lines AG, AZ, AD be equal and those at B, GB, ZB, DB equal too. (See diagram.)

Draw the line AEB. Then the triangles are equal; for their base Aeb is equal. Draw perpendiculars to AEB from the angles; GE from G, Ze from Z, DE from D. Then these perpendiculars are equal, being in equal triangles. And they are all in one plane, being all at right angles to AEB and meeting at a single point E. So if you draw the line it will be a circle and E its centre. Now B is the sun, A the eye, and the circumference passing through the points GZD the cloud from which the line of sight is reflected to the sun.

The mirrors must be thought of as contiguous: each of them is too small to be visible, but their contiguity makes the whole made up of them all to seem one. The bright band is the sun, which is seen as a circle, appearing successively in each of the mirrors as a point indivisible to sense. The band of cloud next to it is black, its colour being intensified by contrast with the brightness of the halo. The halo is formed rather near the earth because that is calmer: for where there is wind it is clear that no halo can maintain its position.

Haloes are commoner round the moon because the greater heat of the sun dissolves the condensations of the air more rapidly.

Haloes are formed round stars for the same reasons, but they are not prognostic in the same way because the condensation they imply is so insignificant as to be barren.

Part 4

We have already stated that the rainbow is a reflection: we have now to explain what sort of reflection it is, to describe its various concomitants, and to assign their causes.

Sight is reflected from all smooth surfaces, such as are air and water among others. Air must be condensed if it is to act as a mirror, though it often gives a reflection even uncondensed when the sight is weak. Such was the case of a man whose sight was faint and indistinct. He always saw an image in front of him and facing him as he walked. This was because his sight was reflected back to him. Its morbid condition made it so weak and delicate that the air close by acted as a mirror, just as distant and condensed air normally does, and his sight could not push it back. So promontories in the sea 'loom' when there is a south-east wind, and everything seems bigger, and in a mist, too, things seem bigger: so, too, the sun and the stars seem bigger when rising and setting than on the meridian. But things are best reflected from water, and even in process of formation it is a better mirror than air, for each of the particles, the union of which constitutes a raindrop, is necessarily a better mirror than mist. Now it is obvious and has already been stated that a mirror of this kind renders the colour of an object only, but not its shape. Hence it follows that when it is on the point of raining and the air in the clouds is in process of forming into raindrops but the rain is not yet actually there, if the sun is opposite, or any other object bright enough to make the cloud a mirror and cause the sight to be reflected to the object then the reflection must render the colour of the object without its shape. Since each of the mirrors is so small as to be invisible and what we see is the continuous magnitude made up of them all, the reflection necessarily gives us a continuous magnitude made up of one colour; each of the mirrors contributing the same colour to the whole. We may deduce that since these conditions are realizable there will be an appearance due to reflection whenever the sun and the cloud are related in the way described and we are between them. But these are just the conditions under which the rainbow appears. So it is clear that the rainbow is a reflection of sight to the sun.

So the rainbow always appears opposite the sun whereas the halo is round it. They are both reflections, but the rainbow is distinguished by the variety of its colours. The reflection in the one case is from water which is dark and from a distance; in the other from air which is nearer and lighter in colour. White light through a dark medium or on a dark surface (it makes no difference) looks red. We know how red the flame of green wood is: this is because so much smoke is mixed with the bright white firelight: so, too, the sun appears red through smoke and mist. That is why in the rainbow reflection the outer circumference is red (the reflection being from small particles of water), but not in the case of the halo. The other colours shall be explained later. Again, a condensation of this kind cannot persist in the neighbourhood of the sun: it must either turn to rain or be dissolved, but opposite to the sun there is an interval during which the water is formed. If there were not this distinction haloes would be coloured like the rainbow. Actually no complete or circular halo presents this colour, only small and fragmentary appearances called 'rods'. But if a haze due to water or any other dark substance formed there we should have had, as we maintain, a complete rainbow like that which we do find lamps. A rainbow appears round these in winter, generally with southerly winds. Persons whose eyes are moist see it most clearly because their sight is weak and easily reflected. It is due to the moistness of the air and the soot which the flame gives off and which mixes with the air and makes it a mirror, and to the blackness which that mirror derives from the smoky nature of the soot. The light of the lamp appears as a circle which is not white but purple. It shows the colours of the rainbow; but because the sight that is reflected is too weak and the mirror too dark, red is absent. The rainbow that is seen when oars are raised out of the sea involves the same relative positions as that in the sky, but its colour is more like that round the lamps, being purple rather than red. The reflection is from very small particles continuous with one another, and in this case the particles are fully formed water. We get a rainbow, too, if a man sprinkles fine drops in a room turned to the sun so that the sun is shining in part of the room and throwing a shadow in the rest. Then if one man sprinkles in the room, another, standing outside, sees a rainbow where the sun's rays cease and make the shadow. Its nature and colour is like that from the oars and its cause is the same, for the sprinkling hand corresponds to the oar.

That the colours of the rainbow are those we described and how the other colours come to appear in it will be clear from the following considerations. We must recognize, as we have said, and lay down: first, that white colour on a black surface or seen through a black medium gives red; second, that sight when strained to a distance becomes weaker and less; third, that black is in a sort the negation of sight: an object is black because sight fails; so everything at a distance looks blacker, because sight does not reach it. The theory of these matters belongs to the account of the senses, which are the proper subjects of such an inquiry; we need only state about them what is necessary for us. At all events, that is the reason why distant objects and objects seen in a mirror look darker and smaller and smoother, why the reflection of clouds in water is darker than the clouds themselves. This latter is clearly the case: the reflection diminishes the sight that reaches them. It makes no difference whether the change is in the object seen or. in the sight, the result being in either case the same. The following fact further is worth noticing. When there is a cloud near the sun and we look at it does not look coloured at all but white, but when we look at the same cloud in water it shows a trace of rainbow colouring. Clearly, then, when sight is reflected it is weakened and, as it makes dark look darker, so it makes white look less white, changing it and bringing it nearer to black. When the sight is relatively strong the change is to red; the next stage is green, and a further degree of weakness gives violet. No further change is visible, but three completes the series of colours (as we find three does in most other things), and the change into the rest is imperceptible to sense. Hence also the rainbow appears with three colours; this is true of each of the two, but in a contrary way. The outer band of the primary rainbow is red: for the largest band reflects most sight to the sun, and the outer band is largest. The middle band and the third go on the same principle. So if the principles we laid down about the appearance of colours are true the rainbow necessarily has three colours, and these three and no others. The appearance of yellow is due to contrast, for the red is whitened by its juxtaposition with green. We can see this from the fact that the rainbow is purest when the cloud is blackest; and then the red shows most yellow. (Yellow in the rainbow comes between red and green.) So the whole of the red shows white by contrast with the blackness of the cloud around: for it is white compared to the cloud and the green. Again, when the rainbow is fading away and the red is dissolving, the white cloud is brought into contact with the green and becomes yellow. But the moon rainbow affords the best instance of this colour contrast. It looks quite white: this is because it appears on the dark cloud and at night. So, just as fire is intensified by added fire, black beside black makes that which is in some degree white look quite white. Bright dyes too show the effect of contrast. In woven and embroidered stuffs the appearance of colours is profoundly affected by their juxtaposition with one another (purple, for instance, appears different on white and on black wool), and also by differences of illumination. Thus embroiderers say that they often make mistakes in their colours when they work by lamplight, and use the wrong ones.

We have now shown why the rainbow has three colours and that these are its only colours. The same cause explains the double rainbow and the faintness of the colours in the outer one and their inverted order. When sight is strained to a great distance the appearance of the distant object is affected in a certain way: and the same thing holds good here. So the reflection from the outer rainbow is weaker because it takes place from a greater distance and less of it reaches the sun, and so the colours seen are fainter. Their order is reversed because more reflection reaches the sun from the smaller, inner band. For that reflection is nearer to our sight which is reflected from the band which is nearest to the primary rainbow. Now the smallest band in the outer rainbow is that which is nearest, and so it will be red; and the second and the third will follow the same principle. Let B be the outer rainbow, A the inner one; let R stand for the red colour, G for green, V for violet; yellow appears at the point Y. Three rainbows or more are not found because even the second is fainter, so that the third reflection can have no strength whatever and cannot reach the sun at all. (See diagram.)

Part 5

The rainbow can never be a circle nor a segment of a circle greater than a semicircle. The consideration of the diagram will prove this and the other properties of the rainbow. (See diagram.)

Let A be a hemisphere resting on the circle of the horizon, let its centre be K and let H be another point appearing on the horizon. Then, if the lines that fall in a cone from K have HK as their axis, and, K and M being joined, the lines KM are reflected from the hemisphere to H over the greater angle, the lines from K will fall on the circumference of a circle. If the reflection takes place when the luminous body is rising or setting the segment of the circle above the earth which is cut off by the horizon will be a semi-circle; if the luminous body is above the horizon it will always be less than a semicircle, and it will be smallest when the luminous body culminates. First let the luminous body be appearing on the horizon at the point H, and let KM be reflected to H, and let the plane in which A is, determined by the triangle HKM, be produced. Then the section of the sphere will be a great circle. Let it be A (for it makes no difference which of the planes passing through the line HK and determined by the triangle KMH is produced). Now the lines drawn from H and K to a point on the semicircle A are in a certain ratio to one another, and no lines drawn from the same points to another point on that semicircle can have the same ratio. For since both the points H and K and the line KH are given, the line MH will be given too; consequently the ratio of the line MH to the line MK will be given too. So M will touch a given circumference. Let this be NM. Then the intersection of the circumferences is given, and the same ratio cannot hold between lines in the same plane drawn from the same points to any other circumference but MN.

Draw a line DB outside of the figure and divide it so that D:B=MH:MK. But MH is greater than MK since the reflection of the cone is over the greater angle (for it subtends the greater angle of the triangle KMH). Therefore D is greater than B. Then add to B a line Z such that B+Z:D=D:B.

Then make another line having the same ratio to B as KH has to Z, and join MI.

Then I is the pole of the circle on which the lines from K fall. For the ratio of D to IM is the same as that of Z to KH and of B to KI. If not, let D be in the same ratio to a line indifferently lesser or greater than IM, and let this line be IP. Then HK and KI and IP will have the same ratios to one another as Z, B, and D. But the ratios between Z, B, and D were such that Z+B:D=D: B. Therefore Ih:IP=IP:IK. Now, if the points K, H be joined with the point P by the lines HP, KP, these lines will be to one another as IH is to IP, for the sides of the triangles HIP, KPI about the angle I are homologous. Therefore, HP too will be to KP as HI is to IP. But this is also the ratio of MH to MK, for the ratio both of HI to IP and of Mh to MK is the same as that of D to B. Therefore, from the points H, K there will have been drawn lines with the same ratio to one another, not only to the circumference MN but to another point as well, which is impossible. Since then D cannot bear that ratio to any line either lesser or greater than IM (the proof being in either case the same), it follows that it must stand in that ratio to MI itself. Therefore as MI is to IK so IH will be to MI and finally MH to Mk.

If, then, a circle be described with I as pole at the distance MI it will touch all the angles which the lines from H and K make by their reflection. If not, it can be shown, as before, that lines drawn to different points in the semicircle will have the same ratio to one another, which was impossible. If, then, the semicircle A be revolved about the diameter HKI, the lines reflected from the points H, K at the point M will have the same ratio, and will make the angle KMH equal, in every plane. Further, the angle which HM and MI make with HI will always be the same. So there are a number of triangles on HI and KI equal to the triangles HMI and KMI. Their perpendiculars will fall on HI at the same point and will be equal. Let O be the point on which they fall. Then O is the centre of the circle, half of which, MN, is cut off by the horizon. (See diagram.)

Next let the horizon be ABG but let H have risen above the horizon. Let the axis now be HI. The proof will be the same for the rest as before, but the pole I of the circle will be below the horizon Ag since the point H has risen above the horizon. But the pole, and the centre of the circle, and the centre of that circle (namely HI) which now determines the position of the sun are on the same line. But since KH lies above the diameter AG, the centre will be at O on the line KI below the plane of the circle AG determined the position of the sun before. So the segment YX which is above the horizon will be less than a semicircle. For YXM was a semicircle and it has now been cut off by the horizon AG. So part of it, YM, will be invisible when the sun has risen above the horizon, and the segment visible will be smallest when the sun is on the meridian; for the higher

H is the lower the pole and the centre of the circle will be.

In the shorter days after the autumn equinox there may be a rainbow at any time of the day, but in the longer days from the spring to the autumn equinox there cannot be a rainbow about midday. The reason for this is that when the sun is north of the equator the visible arcs of its course are all greater than a semicircle, and go on increasing, while the invisible arc is small, but when the sun is south of the equator the visible arc is small and the invisible arc great, and the farther the sun moves south of the equator the greater is the invisible arc. Consequently, in the days near the summer solstice, the size of the visible arc is such that before the point H reaches the middle of that arc, that is its point of culmination, the point is well below the horizon; the reason for this being the great size of the visible arc, and the consequent distance of the point of culmination from the earth. But in the days near the winter solstice the visible arcs are small, and the contrary is necessarily the case: for the sun is on the meridian before the point H has risen far.

Part 6

Mock suns, and rods too, are due to the causes we have described. A mock sun is caused by the reflection of sight to the sun. Rods are seen when sight reaches the sun under circumstances like those which we described, when there are clouds near the sun and sight is reflected from some liquid surface to the cloud. Here the clouds themselves are colourless when you look at them directly, but in the water they are full of rods. The only difference is that in this latter case the colour of the cloud seems to reside in the water, but in the case of rods on the cloud itself. Rods appear when the composition of the cloud is uneven, dense in part and in part rare, and more and less watery in different parts. Then the sight is reflected to the sun: the mirrors are too small for the shape of the sun to appear, but, the bright white light of the sun, to which the sight is reflected, being seen on the uneven mirror, its colour appears partly red, partly green or yellow. It makes no difference whether sight passes through or is reflected from a medium of that kind; the colour is the same in both cases; if it is red in the first case it must be the same in the other.

Rods then are occasioned by the unevenness of the mirror-as regards colour, not form. The mock sun, on the contrary, appears when the air is very uniform, and of the same density throughout.

This is why it is white: the uniform character of the mirror gives the reflection in it a single colour, while the fact that the sight is reflected in a body and is thrown on the sun all together by the mist, which is dense and watery though not yet quite water, causes the sun's true colour to appear just as it does when the reflection is from the dense, smooth surface of copper. So the sun's colour being white, the mock sun is white too. This, too, is the reason why the mock sun is a surer sign of rain than the rods; it indicates, more than they do, that the air is ripe for the production of water. Further a mock sun to the south is a surer sign of rain than one to the north, for the air in the south is readier to turn into water than that in the north.

Mock suns and rods are found, as we stated, about sunset and sunrise, not above the sun nor below it, but beside it. They are not found very close to the sun, nor very far from it, for the sun dissolves the cloud if it is near, but if it is far off the reflection cannot take place, since sight weakens when it is reflected from a small mirror to a very distant object. (This is why a halo is never found opposite to the sun.) If the cloud is above the sun and close to it the sun will dissolve it; if it is above the sun but at a distance the sight is too weak for the reflection to take place, and so it will not reach the sun. But at the side of the sun, it is possible for the mirror to be at such an interval that the sun does not dissolve the cloud, and yet sight reaches it undiminished because it moves close to the earth and is not dissipated in the immensity of space. It cannot subsist below the sun because close to the earth the sun's rays would dissolve it, but if it were high up and the sun in the middle of the heavens, sight would be dissipated. Indeed, even by the side of the sun, it is not found when the sun is in the middle of the sky, for then the line of vision is not close to the earth, and so but little sight reaches the mirror and the reflection from it is altogether feeble.

Some account has now been given of the effects of the secretion above the surface of the earth; we must go on to describe its operations below, when it is shut up in the parts of the earth.

Just as its twofold nature gives rise to various effects in the upper region, so here it causes two varieties of bodies. We maintain that there are two exhalations, one vaporous the other smoky, and there correspond two kinds of bodies that originate in the earth, 'fossiles' and metals. The heat of the dry exhalation is the cause of all 'fossiles'. Such are the kinds of stones that cannot be melted, and realgar, and ochre, and ruddle, and sulphur, and the other things of that kind, most 'fossiles' being either coloured lye or, like cinnabar, a stone compounded of it. The vaporous exhalation is the cause of all metals, those bodies which are either fusible or malleable such as iron, copper, gold. All these originate from the imprisonment of the vaporous exhalation in the earth, and especially in stones. Their dryness compresses it, and it congeals just as dew or hoar-frost does when it has been separated off, though in the present case the metals are generated before that segregation occurs. Hence, they are water in a sense, and in a sense not. Their matter was that which might have become water, but it can no longer do so: nor are they, like savours, due to a qualitative change in actual water. Copper and gold are not formed like that, but in every case the evaporation congealed before water was formed. Hence, they all (except gold) are affected by fire, and they possess an admixture of earth; for they still contain the dry exhalation.

This is the general theory of all these bodies, but we must take up each kind of them and discuss it separately.

BOOK IV

Part 1

WE have explained that the qualities that constitute the elements are four, and that their combinations determine the number of the elements to be four.

Two of the qualities, the hot and the cold, are active; two, the dry and the moist, passive. We can satisfy ourselves of this by looking at instances. In every case heat and cold determine, conjoin, and change things of the same kind and things of different kinds, moistening, drying, hardening, and softening them. Things dry and moist, on the other hand, both in isolation and when present together in the same body are the subjects of that determination and of the other affections enumerated. The account we give of the qualities when we define their character shows this too. Hot and cold we describe as active, for 'congregating' is essentially a species of 'being active': moist and dry are passive, for it is in virtue of its being acted upon in a certain way that a thing is said to be 'easy to determine' or 'difficult to determine'. So it is clear that some of the qualities are active and some passive.

Next we must describe the operations of the active qualities and the forms taken by the passive. First of all, true becoming, that is, natural change, is always the work of these powers and so is the corresponding natural destruction; and this becoming and this destruction are found in plants and animals and their parts. True natural becoming is a change introduced by these powers into the matter underlying a given thing when they are in a certain ratio to that matter, which is the passive qualities we have mentioned. When the hot and the cold are masters of the matter they generate a thing: if they are not, and the failure is partial, the object is imperfectly boiled or otherwise unconcocted. But the strictest general opposite of true becoming is putrefaction. All natural destruction is on the way to it, as are, for instance, growing old or growing dry. Putrescence is the end of all these things, that is of all natural objects, except such as are destroyed by violence: you can burn, for instance, flesh, bone, or anything else, but the natural course of their destruction ends in putrefaction. Hence things that putrefy begin by being moist and end by being dry. For the moist and the dry were their matter, and the operation of the active qualities caused the dry to be determined by the moist.

Destruction supervenes when the determined gets the better of the determining by the help of the environment (though in a special sense the word putrefaction is applied to partial destruction, when a thing's nature is perverted). Hence everything, except fire, is liable to putrefy; for earth, water, and air putrefy, being all of them matter relatively to fire. The definition of putrefaction is: the destruction of the peculiar and natural heat in any moist subject by external heat, that is, by the heat of the environment. So since lack of heat is the ground of this affection and everything in as far as it lacks heat is cold, both heat and cold will be the causes of putrefaction, which will be due indifferently to cold in the putrefying subject or to heat in the environment.

This explains why everything that putrefies grows drier and ends by becoming earth or dung. The subject's own heat departs and causes the natural moisture to evaporate with it, and then there is nothing left to draw in moisture, for it is a thing's peculiar heat that attracts moisture and draws it in. Again, putrefaction takes place less in cold that in hot seasons, for in winter the surrounding air and water contain but little heat and it has no power, but in summer there is more. Again, what is frozen does not putrefy, for its cold is greater that the heat of the air and so is not mastered, whereas what affects a thing does master it. Nor does that which is boiling or hot putrefy, for the heat in the air being less than that in the object does not prevail over it or set up any change. So too anything that is flowing or in motion is less apt to putrefy than a thing at rest, for the motion set up by the heat in the air is weaker than that pre-existing in the object, and so it causes no change. For the same reason a great quantity of a thing putrefies less readily than a little, for the greater quantity contains too much proper fire and cold for the corresponding qualities in the environment to get the better of. Hence, the sea putrefies quickly when broken up into parts, but not as a whole; and all other waters likewise. Animals too are generated in putrefying bodies, because the heat that has been secreted, being natural, organizes the particles secreted with it.

So much for the nature of becoming and of destruction.

Part 2

We must now describe the next kinds of processes which the qualities already mentioned set up in actually existing natural objects as matter.

Of these concoction is due to heat; its species are ripening, boiling, broiling. Inconcoction is due to cold and its species are rawness, imperfect boiling, imperfect broiling. (We must recognize that the things are not properly denoted by these words: the various classes of similar objects have no names universally applicable to them; consequently we must think of the species enumerated as being not what those words denote but something like it.) Let us say what each of them is. Concoction is a process in which the natural and proper heat of an object perfects the corresponding passive qualities, which are the proper matter of any given object. For when concoction has taken place we say that a thing has been perfected and has come to be itself. It is the proper heat of a thing that sets up this perfecting, though external influences may contribute in some degrees to its fulfillment. Baths, for instance, and other things of the kind contribute to the digestion of food, but the primary cause is the proper heat of the body. In some cases of concoction the end of the process is the nature of the thing-nature, that is, in the sense of the formal cause and essence. In other cases it leads to some presupposed state which is attained when the moisture has acquired certain properties or a certain magnitude in the process of being broiled or boiled or of putrefying, or however else it is being heated. This state is the end, for when it has been reached the thing has some use and we say that concoction has taken place. Must is an instance of this, and the matter in boils when it becomes purulent, and tears when they become rheum, and so with the rest.

Concoction ensues whenever the matter, the moisture, is mastered. For the matter is what is determined by the heat connatural to the object, and as long as the ratio between them exists in it a thing maintains its nature. Hence things like the liquid and solid excreta and ejecta in general are signs of health, and concoction is said to have taken place in them, for they show that the proper heat has got the better of the indeterminate matter.

Things that undergo a process of concoction necessarily become thicker and hotter, for the action of heat is to make things more compact, thicker, and drier.

This then is the nature of concoction: but inconcoction is an imperfect state due to lack of proper heat, that is, to cold. That of which the imperfect state is, is the corresponding passive qualities which are the natural matter of anything.

So much for the definition of concoction and inconcoction.

Part 3

Ripening is a sort of concoction; for we call it ripening when there is a concoction of the nutriment in fruit. And since concoction is a sort of perfecting, the process of ripening is perfect when the seeds in fruit are able to reproduce the fruit in which they are found; for in all other cases as well this is what we mean by 'perfect'. This is what 'ripening' means when the word is applied to fruit. However, many other things that have undergone concoction are said to be 'ripe', the general character of the process being the same, though the word is applied by an extension of meaning. The reason for this extension is, as we explained before, that the various modes in which natural heat and cold perfect the matter they determine have not special names appropriated to them. In the case of boils and phlegm, and the like, the process of ripening is the concoction of the moisture in them by their natural heat, for only that which gets the better of matter can determine it. So everything that ripens is condensed from a spirituous into a watery state, and from a watery into an earthy state, and in general from being rare becomes dense. In this process the nature of the thing that is ripening incorporates some of the matter in itself, and some it rejects. So much for the definition of ripening.

Rawness is its opposite and is therefore an imperfect concoction of the nutriment in the fruit, namely, of the undetermined moisture. Consequently a raw thing is either spirituous or watery or contains both spirit and water. Ripening being a kind of perfecting, rawness will be an imperfect state, and this state is due to a lack of natural heat and its disproportion to the moisture that is undergoing the process of ripening. (Nothing moist ripens without the admixture of some dry matter: water alone of liquids does not thicken.) This disproportion may be due either to defect of heat or to excess of the matter to be determined: hence the juice of raw things is thin, cold rather than hot, and unfit for food or drink. Rawness, like ripening, is used to denote a variety of states. Thus the liquid and solid excreta and catarrhs are called raw for the same reason, for in every case the word is applied to things because their heat has not got the mastery in them and compacted them. If we go further, brick is called raw and so is milk and many other things too when they are such as to admit of being changed and compacted by heat but have remained unaffected. Hence, while we speak of 'boiled' water, we cannot speak of raw water, since it does not thicken. We have now defined ripening and rawness and assigned their causes.

Boiling is, in general, a concoction by moist heat of the indeterminate matter contained in the moisture of the thing boiled, and the word is strictly applicable only to things boiled in the way of cooking. The indeterminate matter, as we said, will be either spirituous or watery. The cause of the concoction is the fire contained in the moisture; for what is cooked in a frying-pan is broiled: it is the heat outside that affects it and, as for the moisture in which it is contained, it dries this up and draws it into itself. But a thing that is being boiled behaves in the opposite way: the moisture contained in it is drawn out of it by the heat in the liquid outside. Hence boiled meats are drier than broiled; for, in boiling, things do not draw the moisture into themselves, since the external heat gets the better of the internal: if the internal heat had got the better it would have drawn the moisture to itself. Not every body admits of the process of boiling: if there is no moisture in it, it does not (for instance, stones), nor does it if there is moisture in it but the density of the body is too great for it-to-be mastered, as in the case of wood. But only those bodies can be boiled that contain moisture which can be acted on by the heat contained in the liquid outside. It is true that gold and wood and many other things are said to be 'boiled': but this is a stretch of the meaning of the word, though the kind of thing intended is the same, the reason for the usage being that the various cases have no names appropriated to them. Liquids too, like milk and must, are said to undergo a process of 'boiling' when the external fire that surrounds and heats them changes the savour in the liquid into a given form, the process being thus in a way like what we have called boiling.

The end of the things that undergo boiling, or indeed any form of concoction, is not always the same: some are meant to be eaten, some drunk, and some are intended for other uses; for instance dyes, too, are said to be 'boiled'.

All those things then admit of 'boiling' which can grow denser, smaller, or heavier; also those which do that with a part of themselves and with a part do the opposite, dividing in such a way that one portion thickens while the other grows thinner, like milk when it divides into whey and curd. Oil by itself is affected in none of these ways, and therefore cannot be said to admit of 'boiling'. Such then is the species of concoction known as 'boiling', and the process is the same in an artificial and in a natural instrument, for the cause will be the same in every case.

Imperfect boiling is the form of inconcoction opposed to boiling. Now the opposite of boiling properly so called is an inconcoction of the undetermined matter in a body due to lack of heat in the surrounding liquid. (Lack of heat implies, as we have pointed out, the presence of cold.) The motion which causes imperfect boiling is different from that which causes boiling, for the heat which operates the concoction is driven out. The lack of heat is due either to the amount of cold in the liquid or to the quantity of moisture in the object undergoing the process of boiling. Where either of these conditions is realized the heat in the surrounding liquid is too great to have no effect at all, but too small to carry out the process of concocting uniformly and thoroughly. Hence things are harder when they are imperfectly boiled than when they are boiled, and the moisture in them more distinct from the solid parts. So much for the definition and causes of boiling and imperfect boiling.

Broiling is concoction by dry foreign heat. Hence if a man were to boil a thing but the change and concoction in it were due, not to the heat of the liquid but to that of the fire, the thing will have been broiled and not boiled when the process has been carried to completion: if the process has gone too far we use the word 'scorched' to describe it. If the process leaves the thing drier at the end the agent has been dry heat. Hence the outside is drier than the inside, the opposite being true of things boiled. Where the process is artificial, broiling is more difficult than boiling, for it is difficult to heat the inside and the outside uniformly, since the parts nearer to the fire are the first to get dry and consequently get more intensely dry. In this way the outer pores contract and the moisture in the thing cannot be secreted but is shut in by the closing of the pores. Now broiling and boiling are artificial processes, but the same general kind of thing, as we said, is found in nature too. The affections produced are similar though they lack a name; for art imitates nature. For instance, the concoction of food in the body is like boiling, for it takes place in a hot and moist medium and the agent is the heat of the body. So, too, certain forms of indigestion are like imperfect boiling. And it is not true that animals are generated in the concoction of food, as some say. Really they are generated in the excretion which putrefies in the lower belly, and they ascend afterwards. For concoction goes on in the upper belly but the excretion putrefies in the lower: the reason for this has been explained elsewhere.

We have seen that the opposite of boiling is imperfect boiling: now there is something correspondingly opposed to the species of concoction called broiling, but it is more difficult to find a name for it. It would be the kind of thing that would happen if there were imperfect broiling instead of broiling proper through lack of heat due to deficiency in the external fire or to the quantity of water in the thing undergoing the process. For then we should get too much heat for no effect to be produced, but too little for concoction to take place.

We have now explained concoction and inconcoction, ripening and rawness, boiling and broiling, and their opposites.

Part 4

We must now describe the forms taken by the passive qualities the moist and the dry. The elements of bodies, that is, the passive ones, are the moist and the dry; the bodies themselves are compounded of them and whichever predominates determines the nature of the body; thus some bodies partake more of the dry, others of the moist. All the forms to be described will exist either actually, or potentially and in their opposite: for instance, there is actual melting and on the other hand that which admits of being melted.

Since the moist is easily determined and the dry determined with difficulty, their relation to one another is like that of a dish and its condiments. The moist is what makes the dry determinable, and each serves as a sort of glue to the other-as Empedocles said in his poem on Nature, 'glueing meal together by means of water.' Thus the determined body involves them both. Of the elements earth is especially representative of the dry, water of the moist, and therefore all determinate bodies in our world involve earth and water. Every body shows the quality of that element which predominates in it. It is because earth and water are the material elements of all bodies that animals live in them alone and not in air or fire.

Of the qualities of bodies hardness and softness are those which must primarily belong to a determined thing, for anything made up of the dry and the moist is necessarily either hard or soft. Hard is that the surface of which does not yield into itself; soft that which does yield but not by interchange of place: water, for instance, is not soft, for its surface does not yield to pressure or sink in but there is an interchange of place. Those things are absolutely hard and soft which satisfy the definition absolutely, and those things relatively so which do so compared with another thing. Now relatively to one another hard and soft are indefinable, because it is a matter of degree, but since all the objects of sense are determined by reference to the faculty of sense it is clearly the relation to touch which determines that which is hard and soft absolutely, and touch is that which we use as a standard or mean. So we call that which exceeds it hard and that which falls short of it soft.

Part 5

A body determined by its own boundary must be either hard or soft; for it either yields or does not.

It must also be concrete: or it could not be so determined. So since everything that is determined and solid is either hard or soft and these qualities are due to concretion, all composite and determined bodies must involve concretion. Concretion therefore must be discussed.

Now there are two causes besides matter, the agent and the quality brought about, the agent being the efficient cause, the quality the formal cause. Hence concretion and disaggregation, drying and moistening, must have these two causes.

But since concretion is a form of drying let us speak of the latter first.

As we have explained, the agent operates by means of two qualities and the patient is acted on in virtue of two qualities: action takes place by means of heat or cold, and the quality is produced either by the presence or by the absence of heat or cold; but that which is acted upon is moist or dry or a compound of both. Water is the element characterized by the moist, earth that characterized by the dry, for these among the elements that admit the qualities moist and dry are passive. Therefore cold, too, being found in water and earth (both of which we recognize to be cold), must be reckoned rather as a passive quality. It is active only as contributing to destruction or incidentally in the manner described before; for cold is sometimes actually said to burn and to warm, but not in the same way as heat does, but by collecting and concentrating heat.

The subjects of drying are water and the various watery fluids and those bodies which contain water either foreign or connatural. By foreign I mean like the water in wool, by connatural, like that in milk. The watery fluids are wine, urine, whey, and in general those fluids which have no sediment or only a little, except where this absence of sediment is due to viscosity. For in some cases, in oil and pitch for instance, it is the viscosity which prevents any sediment from appearing.

It is always a process of heating or cooling that dries things, but the agent in both cases is heat, either internal or external. For even when things are dried by cooling, like a garment, where the moisture exists separately it is the internal heat that dries them. It carries off the moisture in the shape of vapour (if there is not too much of it), being itself driven out by the surrounding cold. So everything is dried, as we have said, by a process either of heating or cooling, but the agent is always heat, either internal or external, carrying off the moisture in vapour. By external heat I mean as where things are boiled: by internal where the heat breathes out and takes away and uses up its moisture. So much for drying.

Part 6

Liquefaction is, first, condensation into water; second, the melting of a solidified body. The first, condensation, is due to the cooling of vapour: what melting is will appear from the account of solidification.

Whatever solidifies is either water or a mixture of earth and water, and the agent is either dry heat or cold. Hence those of the bodies solidified by heat or cold which are soluble at all are dissolved by their opposites. Bodies solidified by the dry-hot are dissolved by water, which is the moist-cold, while bodies solidified by cold are dissolved by fire, which is hot. Some things seem to be solidified by water, e.g. boiled honey, but really it is not the water but the cold in the water which effects the solidification. Aqueous bodies are not solidified by fire: for it is fire that dissolves them, and the same cause in the same relation cannot have opposite effects upon the same thing. Again, water solidifies owing to the departure of heat; so it will clearly be dissolved by the entry into it of heat: cold, therefore, must be the agent in solidifying it.

Hence aqueous bodies do not thicken when they solidify; for thickening occurs when the moisture goes off and the dry matter comes together, but water is the only liquid that does not thicken. Those bodies that are made up of both earth and water are solidified both by fire and by cold and in either case are thickened. The operation of the two is in a way the same and in a way different. Heat acts by drawing off the moisture, and as the moisture goes off in vapour the dry matter thickens and collects. Cold acts by driving out the heat, which is accompanied by the moisture as this goes off in vapour with it. Bodies that are soft but not liquid do not thicken but solidify when the moisture leaves them, e.g. potter's clay in process of baking: but those mixed bodies that are liquid thicken besides solidifying, like milk. Those bodies which have first been thickened or hardened by cold often begin by becoming moist: thus potter's clay at first in the process of baking steams and grows softer, and is liable to distortion in the ovens for that reason.

Now of the bodies solidified by cold which are made up both of earth and water but in which the earth preponderates, those which solidify by the departure of heat melt by heat when it enters into them again; this is the case with frozen mud. But those which solidify by refrigeration, where all the moisture has gone off in vapour with the heat, like iron and horn, cannot be dissolved except by excessive heat, but they can be softened-though manufactured iron does melt, to the point of becoming fluid and then solidifying again. This is how steel is made. The dross sinks to the bottom and is purged away: when this has been done often and the metal is pure we have steel. The process is not repeated often because the purification of the metal involves great waste and loss of weight. But the iron that has less dross is the better iron. The stone pyrimachus, too, melts and forms into drops and becomes fluid; after having been in a fluid state it solidifies and becomes hard again. Millstones, too, melt and become fluid: when the fluid mass begins to solidify it is black but its consistency comes to be like that of lime. and earth, too

Of the bodies which are solidified by dry heat some are insoluble, others are dissolved by liquid.

Pottery and some kinds of stone that are formed out of earth burnt up by fire, such as millstones, cannot be dissolved. Natron and salt are soluble by liquid, but not all liquid but only such as is cold. Hence water and any of its varieties melt them, but oil does not. For the opposite of the dry-hot is the cold-moist and what the one solidified the other will dissolve, and so opposites will have opposite effects.

Part 7

If a body contains more water than earth fire only thickens it: if it contains more earth fire solidifies it. Hence natron and salt and stone and potter's clay must contain more earth.

The nature of oil presents the greatest problem. If water preponderated in it, cold ought to solidify it; if earth preponderated, then fire ought to do so. Actually neither solidifies, but both thicken it. The reason is that it is full of air (hence it floats on the top of water, since air tends to rise). Cold thickens it by turning the air in it into water, for any mixture of oil and water is thicker than either. Fire and the lapse of time thicken and whiten it. The whitening follows on the evaporation of any water that may have been in it; the is due to the change of the air into water as the heat in the oil is dissipated. The effect in both cases is the same and the cause is the same, but the manner of its operation is different. Both heat and cold thicken it, but neither dries it (neither the sun nor cold dries oil), not only because it is glutinous but because it contains air. Its glutinous nature prevents it from giving off vapour and so fire does not dry it or boil it off.

Those bodies which are made up of earth and water may be classified according to the preponderance of either. There is a kind of wine, for instance, which both solidifies and thickens by boiling-I mean, must. All bodies of this kind lose their water as they That it is their water may be seen from the fact that the vapour from them condenses into water when collected. So wherever some sediment is left this is of the nature of earth. Some of these bodies, as we have said, are also thickened and dried by cold. For cold not only solidifies but also dries water, and thickens things by turning air into water. (Solidifying, as we have said, is a form of drying.) Now those things that are not thickened by cold, but solidified, belong rather to water, e.g.. wine, urine, vinegar, lye, whey. But those things that are thickened (not by evaporation due to fire) are made up either of earth or of water and air: honey of earth, while oil contains air. Milk and blood, too, are made up of both water and earth, though earth generally predominates in them. So, too, are the liquids out of which natron and salt are formed; and stones are also formed from some mixtures of this kind. Hence, if the whey has not been separated, it burns away if you boil it over a fire. But the earthy element in milk can also be coagulated by the help of fig-juice, if you boil it in a certain way as doctors do when they treat it with fig-juice, and this is how the whey and the cheese are commonly separated. Whey, once separated, does not thicken, as the milk did, but boils away like water. Sometimes, however, there is little or no cheese in milk, and such milk is not nutritive and is more like water. The case of blood is similar: cold dries and so solidifies it. Those kinds of blood that do not solidify, like that of the stag, belong rather to water and are very cold. Hence they contain no fibres: for the fibres are of earth and solid, and blood from which they have been removed does not solidify. This is because it cannot dry; for what remains is water, just as what remains of milk when cheese has been removed is water. The fact that diseased blood will not solidify is evidence of the same thing, for such blood is of the nature of serum and that is phlegm and water, the nature of the animal having failed to get the better of it and digest it.

Some of these bodies are soluble, e.g. natron, some insoluble, e.g. pottery: of the latter, some, like horn, can be softened by heat, others, like pottery and stone, cannot. The reason is that opposite causes have opposite effects: consequently, if solidification is due to two causes, the cold and the dry, solution must be due to the hot and the moist, that is, to fire and to water (these being opposites): water dissolving what was solidified by fire alone, fire what was solidified by cold alone. Consequently, if any things happen to be solidified by the action of both, these are least apt to be soluble. Such a case we find where things have been heated and are then solidified by cold. When the heat in leaving them has caused most of the moisture to evaporate, the cold so compacts these bodies together again as to leave no entrance even for moisture. Therefore heat does not dissolve them (for it only dissolves those bodies that are solidified by cold alone), nor does water (for it does not dissolve what cold solidifies, but only what is solidified by dry heat). But iron is melted by heat and solidified by cold. Wood consists of earth and air and is therefore combustible but cannot be melted or softened by heat. (For the same reason it floats in water-all except ebony. This does not, for other kinds of wood contain a preponderance of air, but in black ebony the air has escaped and so earth preponderates in it.) Pottery consists of earth alone because it solidified gradually in the process of drying. Water cannot get into it, for the pores were only large enough to admit of vapour escaping: and seeing that fire solidified it, that cannot dissolve it either.

So solidification and melting, their causes, and the kinds of subjects in which they occur have been described.

Part 8

All this makes it clear that bodies are formed by heat and cold and that these agents operate by thickening and solidifying. It is because these qualities fashion bodies that we find heat in all of them, and in some cold in so far as heat is absent. These qualities, then, are present as active, and the moist and the dry as passive, and consequently all four are found in mixed bodies. So water and earth are the constituents of homogeneous bodies both in plants and in animals and of metals such as gold, silver, and the rest-water and earth and their respective exhalations shut up in the compound bodies, as we have explained elsewhere.

All these mixed bodies are distinguished from one another, firstly by the qualities special to the various senses, that is, by their capacities of action. (For a thing is white, fragrant, sonant, sweet, hot, cold in virtue of a power of acting on sense). Secondly by other more characteristic affections which express their aptitude to be affected: I mean, for instance, the aptitude to melt or solidify or bend and so forth, all these qualities, like moist and dry, being passive. These are the qualities that differentiate bone, flesh, sinew, wood, bark, stone and all other homogeneous natural bodies. Let us begin by enumerating these qualities expressing the aptitude or inaptitude of a thing to be affected in a certain way. They are as follows: to be apt or inapt to solidify, melt, be softened by heat, be softened by water, bend, break, be comminuted, impressed, moulded, squeezed; to be tractile or non-tractile, malleable or non-malleable, to be fissile or non-fissile, apt or inapt to be cut; to be viscous or friable, compressible or incompressible, combustible or incombustible; to be apt or inapt to give off fumes. These affections differentiate most bodies from one another. Let us go on to explain the nature of each of them. We have already given a general account of that which is apt or inapt to solidify or to melt, but let us return to them again now. Of all the bodies that admit of solidification and hardening, some are brought into this state by heat, others by cold. Heat does this by drying up their moisture, cold by driving out their heat. Consequently some bodies are affected in this way by defect of moisture, some by defect of heat: watery bodies by defect of heat, earthy bodies of moisture. Now those bodies that are so affected by defect of moisture are dissolved by water, unless like pottery they have so contracted that their pores are too small for the particles of water to enter. All those bodies in which this is not the case are dissolved by water, e.g. natron, salt, dry mud. Those bodies that solidified through defect of heat are melted by heat, e.g. ice, lead, copper. So much for the bodies that admit of solidification and of melting, and those that do not admit of melting.

The bodies which do not admit of solidification are those which contain no aqueous moisture and are not watery, but in which heat and earth preponderate, like honey and must (for these are in a sort of state of effervescence), and those which do possess some water but have a preponderance of air, like oil and quicksilver, and all viscous substances such as pitch and birdlime.

Part 9

Those bodies admit of softening which are not (like ice) made up of water, but in which earth predominates. All their moisture must not have left them (as in the case of natron and salt), nor must the relation of dry to moist in them be incongruous (as in the case of pottery). They must be tractile (without admitting water) or malleable (without consisting of water), and the agent in softening them is fire. Such are iron and horn.

Both of bodies that can melt and of bodies that cannot, some do and some do not admit of softening in water. Copper, for instance, which can be melted, cannot be softened in water, whereas wool and earth can be softened in water, for they can be soaked. (It is true that though copper can be melted the agent in its case is not water, but some of the bodies that can be melted by water too such as natron and salt cannot be softened in water: for nothing is said to be so affected unless the water soaks into it and makes it softer.) Some things, on the other hand, such as wool and grain, can be softened by water though they cannot be melted. Any body that is to be softened by water must be of earth and must have its pores larger than the particles of water, and the pores themselves must be able to resist the action of water, whereas bodies that can be 'melted' by water must have pores throughout.

(Why is it that earth is both 'melted' and softened by moisture, while natron is 'melted' but not softened? Because natron is pervaded throughout by pores so that the parts are immediately divided by the water, but earth has also pores which do not connect and is therefore differently affected according as the water enters by one or the other set of pores.)

Some bodies can be bent or straightened, like the reed or the withy, some cannot, like pottery and stone. Those bodies are apt to be bent and straightened which can change from being curved to being straight and from being straight to being curved, and bending and straightening consist in the change or motion to the straight or to a curve, for a thing is said to be in process of being bent whether it is being made to assume a convex or a concave shape. So bending is defined as motion to the convex or the concave without a change of length. For if we added 'or to the straight', we should have a thing bent and straight at once, and it is impossible for that which is straight to be bent. And if all bending is a bending back or a bending down, the former being a change to the convex, the latter to the concave, a motion that leads to the straight cannot be called bending, but bending and straightening are two different things. These, then, are the things that can, and those that cannot be bent, and be straightened.

Some things can be both broken and comminuted, others admit only one or the other. Wood, for instance, can be broken but not comminuted, ice and stone can be comminuted but not broken, while pottery may either be comminuted or broken. The distinction is this: breaking is a division and separation into large parts, comminution into parts of any size, but there must be more of them than two. Now those solids that have many pores not communicating with one another are comminuible (for the limit to their subdivision is set by the pores), but those whose pores stretch continuously for a long way are breakable, while those which have pores of both kinds are both comminuible and breakable.

Some things, e.g. copper and wax, are impressible, others, e.g. pottery and water, are not. The process of being impressed is the sinking of a part of the surface of a thing in response to pressure or a blow, in general to contact. Such bodies are either soft, like wax, where part of the surface is depressed while the rest remains, or hard, like copper. Non-impressible bodies are either hard, like pottery (its surface does not give way and sink in), or liquid, like water (for though water does give way it is not in a part of it, for there is a reciprocal change of place of all its parts). Those impressibles that retain the shape impressed on them and are easily moulded by the hand are called 'plastic'; those that are not easily moulded, such as stone or wood, or are easily moulded but do not retain the shape impressed, like wool or a sponge, are not plastic. The last group are said to be 'squeezable'. Things are 'squeezable' when they can contract into themselves under pressure, their surface sinking in without being broken and without the parts interchanging position as happens in the case of water. (We speak of pressure when there is movement and the motor remains in contact with the thing moved, of impact when the movement is due to the local movement of the motor.) Those bodies are subject to squeezing which have empty pores-empty, that is, of the stuff of which the body itself consists-and that can sink upon the void spaces within them, or rather upon their pores. For sometimes the pores upon which a body sinks in are not empty (a wet sponge, for instance, has its pores full). But the pores, if full, must be full of something softer than the body itself which is to contract. Examples of things squeezable are the sponge, wax, flesh. Those things are not squeezable which cannot be made to contract upon their own pores by pressure, either because they have no pores or because their pores are full of something too hard. Thus iron, stone, water and all liquids are incapable of being squeezed.

Things are tractile when their surface can be made to elongate, for being drawn out is a movement of the surface, remaining unbroken, in the direction of the mover. Some things are tractile, e.g. hair, thongs, sinew, dough, birdlime, and some are not, e.g. water, stone. Some things are both tractile and squeezable, e.g. wool; in other cases the two qualities do not coincide; phlegm, for instance, is tractile but not squeezable, and a sponge squeezable but not tractile.

Some things are malleable, like copper. Some are not, like stone and wood. Things are malleable when their surface can be made to move (but only in part) both downwards and sideways with one and the same blow: when this is not possible a body is not malleable. All malleable bodies are impressible, but not all impressible bodies are malleable, e.g. wood, though on the whole the two go together. Of squeezable things some are malleable and some not: wax and mud are malleable, wool is not. Some things are fissile, e.g. wood, some are not, e.g. potter's clay. A thing is fissile when it is apt to divide in advance of the instrument dividing it, for a body is said to split when it divides to a further point than that to which the dividing instrument divides it and the act of division advances: which is not the case with cutting. Those bodies which cannot behave like this are non-fissile. Nothing soft is fissile (by soft I mean absolutely soft and not relatively: for iron itself may be relatively soft); nor are all hard things fissile, but only such as are neither liquid nor impressible nor comminuible. Such are the bodies that have the pores along which they cohere lengthwise and not crosswise.

Those hard or soft solids are apt to be cut which do not necessarily either split in advance of the instrument or break into minute fragments when they are being divided. Those that necessarily do so and liquids cannot be cut. Some things can be both split and cut, like wood, though generally it is lengthwise that a thing can be split and crosswise that it can be cut. For, a body being divided into many parts fin so far as its unity is made up of many lengths it is apt to be split, in so far as it is made up of many breadths it is apt to be cut.

A thing is viscous when, being moist or soft, it is tractile. Bodies owe this property to the interlocking of their parts when they are composed like chains, for then they can be drawn out to a great length and contracted again. Bodies that are not like this are friable. Bodies are compressible when they are squeezable and retain the shape they have been squeezed into; incompressible when they are either inapt to be squeezed at all or do not retain the shape they have been squeezed into.

Some bodies are combustible and some are not. Wood, wool, bone are combustible; stone, ice are not. Bodies are combustible when their pores are such as to admit fire and their longitudinal pores contain moisture weaker than fire. If they have no moisture, or if, as in ice or very green wood, the moisture is stronger than fire, they are not combustible.

Those bodies give off fumes which contain moisture, but in such a form that it does not go off separately in vapour when they are exposed to fire. For vapour is a moist secretion tending to the nature of air produced from a liquid by the agency of burning heat. Bodies that give off fumes give off secretions of the nature of air by the lapse of time: as they perish away they dry up or become earth. But the kind of secretion we are concerned with now differs from others in that it is not moist nor does it become wind (which is a continuous flow of air in a given direction). Fumes are common secretion of dry and moist together caused by the agency of burning heat. Hence they do not moisten things but rather colour them.

The fumes of a woody body are called smoke. (I mean to include bones and hair and everything of this kind in the same class. For there is no name common to all the objects that I mean, but, for all that, these things are all in the same class by analogy. Compare what Empedocles says: They are one and the same, hair and leaves and the thick wings of birds and scales that grow on stout limbs.) The fumes of fat are a sooty smoke and those of oily substances a greasy steam. Oil does not boil away or thicken by evaporation because it does not give off vapour but fumes. Water on the other hand does not give off fumes, but vapour. Sweet wine does give off fumes, for it contains fat and behaves like oil. It does not solidify under the influence of cold and it is apt to burn. Really it is not wine at all in spite of its name: for it does not taste like wine and consequently does not inebriate as ordinary wine does. It contains but little fumigable stuff and consequently is inflammable.

All bodies are combustible that dissolve into ashes, and all bodies do this that solidify under the influence either of heat or of both heat and cold; for we find that all these bodies are mastered by fire. Of stones the precious stone called carbuncle is least amenable to fire.

Of combustible bodies some are inflammable and some are not, and some of the former are reduced to coals. Those are called 'inflammable' which produce flame and those which do not are called 'non-inflammable'. Those fumigable bodies that are not liquid are inflammable, but pitch, oil, wax are inflammable in conjunction with other bodies rather than by themselves. Most inflammable are those bodies that give off smoke. Of bodies of this kind those that contain more earth than smoke are apt to be reduced to coals. Some bodies that can be melted are not inflammable, e.g. copper; and some bodies that cannot be melted are inflammable, e.g. wood; and some bodies can be melted and are also inflammable, e.g. frankincense. The reason is that wood has its moisture all together and this is continuous throughout and so it burns up: whereas copper has it in each part but not continuous, and insufficient in quantity to give rise to flame. In frankincense it is disposed in both of these ways. Fumigable bodies are inflammable when earth predominates in them and they are consequently such as to be unable to melt. These are inflammable because they are dry like fire. When this dry comes to be hot there is fire. This is why flame is burning smoke or dry exhalation. The fumes of wood are smoke, those of wax and frankincense and such-like, and pitch and whatever contains pitch or such-like are sooty smoke, while the fumes of oil and oily substances are a greasy steam; so are those of all substances which are not at all combustible by themselves because there is too little of the dry in them (the dry being the means by which the transition to fire is effected), but burn very readily in conjunction with something else. (For the fat is just the conjunction of the oily with the dry.) So those bodies that give off fumes, like oil and pitch, belong rather to the moist, but those that burn to the dry.

Part 10

Homogeneous bodies differ to touch-by these affections and differences, as we have said. They also differ in respect of their smell, taste, and colour.

By homogeneous bodies I mean, for instance, 'metals', gold, copper, silver, tin, iron, stone, and everything else of this kind and the bodies that are extracted from them; also the substances found in animals and plants, for instance, flesh, bones, sinew, skin, viscera, hair, fibres, veins (these are the elements of which the non-homogeneous bodies like the face, a hand, a foot, and everything of that kind are made up), and in plants, wood, bark, leaves, roots, and the rest like them.

The homogeneous bodies, it is true, are constituted by a different cause, but the matter of which they are composed is the dry and the moist, that is, water and earth (for these bodies exhibit those qualities most clearly). The agents are the hot and the cold, for they constitute and make concrete the homogeneous bodies out of earth and water as matter. Let us consider, then, which of the homogeneous bodies are made of earth and which of water, and which of both.

Of organized bodies some are liquid, some soft, some hard. The soft and the hard are constituted by a process of solidification, as we have already explained.

Those liquids that go off in vapour are made of water, those that do not are either of the nature of earth, or a mixture either of earth and water, like milk, or of earth and air, like wood, or of water and air, like oil. Those liquids which are thickened by heat are a mixture. (Wine is a liquid which raises a difficulty: for it is both liable to evaporation and it also thickens; for instance new wine does. The reason is that the word 'wine' is ambiguous and different 'wines' behave in different ways. New wine is more earthy than old, and for this reason it is more apt to be thickened by heat and less apt to be congealed by cold. For it contains much heat and a great proportion of earth, as in Arcadia, where it is so dried up in its skins by the smoke that you scrape it to drink. If all wine has some sediment in it then it will belong to earth or to water according to the quantity of the sediment it possesses.) The liquids that are thickened by cold are of the nature of earth; those that are thickened either by heat or by cold consist of more than one element, like oil and honey, and 'sweet wine'.

Of solid bodies those that have been solidified by cold are of water, e.g. ice, snow, hail, hoar-frost. Those solidified by heat are of earth, e.g. pottery, cheese, natron, salt. Some bodies are solidified by both heat and cold. Of this kind are those solidified by refrigeration, that is by the privation both of heat and of the moisture which departs with the heat. For salt and the bodies that are purely of earth solidify by the privation of moisture only, ice by that of heat only, these bodies by that of both. So both the active qualities and both kinds of matter were involved in the process. Of these bodies those from which all the moisture has gone are all of them of earth, like pottery or amber. (For amber, also, and the bodies called 'tears' are formed by refrigeration, like myrrh, frankincense, gum. Amber, too, appears to belong to this class of things: the animals enclosed in it show that it is formed by solidification. The heat is driven out of it by the cold of the river and causes the moisture to evaporate with it, as in the case of honey when it has been heated and is immersed in water.) Some of these bodies cannot be melted or softened; for instance, amber and certain stones, e.g. the stalactites in caves. (For these stalactites, too, are formed in the same way: the agent is not fire, but cold which drives out the heat, which, as it leaves the body, draws out the moisture with it: in the other class of bodies the agent is external fire.) In those from which the moisture has not wholly gone earth still preponderates, but they admit of softening by heat, e.g. iron and horn.

Now since we must include among 'meltables' those bodies which are melted by fire, these contain some water: indeed some of them, like wax, are common to earth and water alike. But those that are melted by water are of earth. Those that are not melted either by fire or water are of earth, or of earth and water.

Since, then, all bodies are either liquid or solid, and since the things that display the affections we have enumerated belong to these two classes and there is nothing intermediate, it follows that we have given a complete account of the criteria for distinguishing whether a body consists of earth or of water or of more elements than one, and whether fire was the agent in its formation, or cold, or both.

Gold, then, and silver and copper and tin and lead and glass and many nameless stone are of water: for they are all melted by heat. Of water, too, are some wines and urine and vinegar and lye and whey and serum: for they are all congealed by cold. In iron, horn, nails, bones, sinews, wood, hair, leaves, bark, earth preponderates. So, too, in amber, myrrh, frankincense, and all the substances called 'tears', and stalactites, and fruits, such as leguminous plants and corn. For things of this kind are, to a greater or less degree, of earth. For of all these bodies some admit of softening by heat, the rest give off fumes and are formed by refrigeration. So again in natron, salt, and those kinds of stones that are not formed by refrigeration and cannot be melted. Blood, on the other hand, and semen, are made up of earth and water and air. If the blood contains fibres, earth preponderates in it: consequently its solidifies by refrigeration and is melted by liquids; if not, it is of water and therefore does not solidify. Semen solidifies by refrigeration, its moisture leaving it together with its heat.

Part 11

We must investigate in the light of the results we have arrived at what solid or liquid bodies are hot and what cold.

Bodies consisting of water are commonly cold, unless (like lye, urine, wine) they contain foreign heat. Bodies consisting of earth, on the other hand, are commonly hot because heat was active in forming them: for instance lime and ashes.

We must recognize that cold is in a sense the matter of bodies. For the dry and the moist are matter (being passive) and earth and water are the elements that primarily embody them, and they are characterized by cold. Consequently cold must predominate in every body that consists of one or other of the elements simply, unless such a body contains foreign heat as water does when it boils or when it has been strained through ashes. This latter, too, has acquired heat from the ashes, for everything that has been burnt contains more or less heat. This explains the generation of animals in putrefying bodies: the putrefying body contains the heat which destroyed its proper heat.

Bodies made up of earth and water are hot, for most of them derive their existence from concoction and heat, though some, like the waste products of the body, are products of putrefaction. Thus blood, semen, marrow, fig juice, and all things of the kinds are hot as long as they are in their natural state, but when they perish and fall away from that state they are so no longer. For what is left of them is their matter and that is earth and water. Hence both views are held about them, some people maintaining them to be cold and others to be warm; for they are observed to be hot when they are in their natural state, but to solidify when they have fallen away from it. That, then, is the case of mixed bodies. However, the distinction we laid down holds good: if its matter is predominantly water a body is cold (water being the complete opposite of fire), but if earth or air it tends to be warm.

It sometimes happens that the coldest bodies can be raised to the highest temperature by foreign heat; for the most solid and the hardest bodies are coldest when deprived of heat and most burning after exposure to fire: thus water is more burning than smoke and stone than water.

Part 12

Having explained all this we must describe the nature of flesh, bone, and the other homogeneous bodies severally.

Our account of the formation of the homogeneous bodies has given us the elements out of which they are compounded and the classes into which they fall, and has made it clear to which class each of those bodies belongs. The homogeneous bodies are made up of the elements, and all the works of nature in turn of the homogeneous bodies as matter. All the homogeneous bodies consist of the elements described, as matter, but their essential nature is determined by their definition. This fact is always clearer in the case of the later products of those, in fact, that are instruments, as it were, and have an end: it is clearer, for instance, that a dead man is a man only in name. And so the hand of a dead man, too, will in the same way be a hand in name only, just as stone flutes might still be called flutes: for these members, too, are instruments of a kind. But in the case of flesh and bone the fact is not so clear to see, and in that of fire and water even less. For the end is least obvious there where matter predominates most. If you take the extremes, matter is pure matter and the essence is pure definition; but the bodies intermediate between the two are matter or definition in proportion as they are near to either. For each of those elements has an end and is not water or fire in any and every condition of itself, just as flesh is not flesh nor viscera viscera, and the same is true in a higher degree with face and hand. What a thing is always determined by its function: a thing really is itself when it can perform its function; an eye, for instance, when it can see. When a thing cannot do so it is that thing only in name, like a dead eye or one made of stone, just as a wooden saw is no more a saw than one in a picture. The same, then, is true of flesh, except that its function is less clear than that of the tongue. So, too, with fire; but its function is perhaps even harder to specify by physical inquiry than that of flesh. The parts of plants, and inanimate bodies like copper and silver, are in the same case. They all are what they are in virtue of a certain power of action or passion-just like flesh and sinew. But we cannot state their form accurately, and so it is not easy to tell when they are really there and when they are not unless the body is thoroughly corrupted and its shape only remains. So ancient corpses suddenly become ashes in the grave and very old fruit preserves its shape only but not its taste: so, too, with the solids that form from milk.

Now heat and cold and the motions they set up as the bodies are solidified by the hot and the cold are sufficient to form all such parts as are the homogeneous bodies, flesh, bone, hair, sinew, and the rest. For they are all of them differentiated by the various qualities enumerated above, tension, tractility, comminuibility, hardness, softness, and the rest of them: all of which are derived from the hot and the cold and the mixture of their motions. But no one would go as far as to consider them sufficient in the case of the non-homogeneous parts (like the head, the hand, or the foot) which these homogeneous parts go to make up. Cold and heat and their motion would be admitted to account for the formation of copper or silver, but not for that of a saw, a bowl, or a box. So here, save that in the examples given the cause is art, but in the nonhomogeneous bodies nature or some other cause.

Since, then, we know to what element each of the homogeneous bodies belongs, we must now find the definition of each of them, the answer, that is, to the question, 'what is' flesh, semen, and the rest? For we know the cause of a thing and its definition when we know the material or the formal or, better, both the material and the formal conditions of its generation and destruction, and the efficient cause of it.

After the homogeneous bodies have been explained we must consider the non-homogeneous too, and lastly the bodies made up of these, such as man, plants, and the rest.

25245705R00053

Printed in Great Britain
by Amazon